F. H. KING

FARMERS

OF

FORTY CENTURIES

OR

PERMANENT AGRICULTURE IN CHINA, KOREA AND JAPAN

By

F. H. KING, D. Sc.

Formerly Professor of Agricultural Physics in the University of Wisconsin
and
Chief of Divison of Soil Management, U. S. Department of Agriculture

Author of "The Soil": "Irrigation and Drainage"; "Physics of
Agriculture" and "Ventilation for Dwellings,
Rural Schools and Stables."

Rodale Press, Inc.
Book Division
Emmaus, Pennsylvania 18049

International Standard Book Number 0-87857-054-3
Library of Congress Catalog Card Number 72-90823

Printed in the United States of America
286

PREFACE

By Dr. L. H. Bailey.

We have not yet gathered up the experience of mankind in the tilling of the earth; yet the tilling of the earth is the bottom condition of civilization. If we are to assemble all the forces and agencies that make for the final conquest of the planet, we must assuredly know how it is that all the peoples in all the places have met the problem of producing their sustenance out of the soil.

We have had few great agricultural travelers and few books that describe the real and significant rural conditions. Of natural history travel we have had very much; and of accounts of sights and events perhaps we have had too many. There are, to be sure, famous books of study and travel in rural regions, and some of them, as Arthur Young's "Travels in France," have touched social and political history; but for the most part, authorship of agricultural travel is yet undeveloped. The spirit of scientific inquiry must now be taken into this field, and all earth-conquest must be compared and the results be given to the people that work.

This was the point of view in which I read Professor King's manuscript. It is the writing of a well-trained observer who went forth not to find diversion or to depict scenery and common wonders, but to study the actual conditions of life of agricultural peoples. We in North America are wont to think that we may instruct all the world in agriculture, because our agricultural wealth is great and our exports to less favored peoples have been heavy; but this wealth is great because our soil is fertile and new, and in large acreage for every person. We have really only begun to farm well. The first condition of farming is to maintain fertility. This condition the oriental peoples have met, and they have solved it in their way. We may never adopt particular methods, but we can profit vastly by their experience. With the increase of personal wants in recent time, the newer countries may never reach such density of population as have Japan and China; but we must nevertheless learn the first lesson in the conservation of natural resources, which are the resources of the land. This is the message that Professor King brought home from the East.

This book on agriculture should have good effect in establishing understanding between the West and the East. If there could be such an interchange of courtesies and inquiries on these themes as is suggested by Professor King, as well as the interchange of athletics and diplomacy and commerce, the common productive people on both sides should gain much that they could use; and the results in amity should be incalculable.

It is a misfortune that Professor King could not have lived to write the concluding ''Message of China and Japan to the World.'' It would have been a careful and forceful summary of his study of eastern conditions. At the moment when the work was going to the printer, he was called suddenly to the endless journey and his travel here was left incomplete. But he bequeathed us a new piece of literature, to add to his standard writings on soils and on the applications of physics and devices to agriculture. Whatever he touched he illuminated.

L. H. BAILEY.

CONTENTS

LIST OF ILLUSTRATIONS

INTRODUCTION.

A word of introduction is needed to place the reader at the best view point from which to consider what is said in the following pages regarding the agricultural practices and customs of China, Korea and Japan. It should be borne in mind that the great factors which today characterize, dominate and determine the agricultural and other industrial operations of western nations were physical impossibilities to them one hundred years ago, and until then had been so to all people.

It should be observed, too, that the United States as yet is a nation of but few people widely scattered over a broad virgin land with more than twenty acres to the support of every man, woman and child, while the people whose practices are to be considered are toiling in fields tilled more than three thousand years and who have scarcely more than two acres per capita,* more than one-half of which is uncultivable mountain land.

Again, the great movement of cargoes of feeding stuffs and mineral fertilizers to western Europe and to the eastern United States began less than a century ago and has never been possible as a means of maintaining soil fertility in China, Korea or Japan, nor can it be continued indefinitely in either Europe or America. These importations are for the time making tolerable the waste of plant food materials through our modern systems of sewage disposal and other faulty practices; but the Mongolian races have held all such wastes, both urban and rural, and many others which we ignore, sacred to agriculture, applying them to their fields.

---

*This figure was wrongly stated in the first edition, as one acre, owing to a mistake in confusing the area of cultivated land with total area.

We are to consider some of the practices of a virile race of some five hundred millions of people who have an unimpaired inheritance moving with the momentum acquired through four thousand years; a people morally and intellectually strong, mechanically capable, who are awakening to a utilization of all the possibilities which science and invention during recent years have brought to western nations; and a people who have long dearly loved peace but who can and will fight in self defense if compelled to do so.

We had long desired to stand face to face with Chinese and Japanese farmers; to walk through their fields and to learn by seeing some of their methods, appliances and practices which centuries of stress and experience have led these oldest farmers in the world to adopt. We desired to learn how it is possible, after twenty and perhaps thirty or even forty centuries, for their soils to be made to produce sufficiently for the maintenance of such dense populations as are living now in these three countries. We have now had this opportunity and almost every day we were instructed, surprised and amazed at the conditions and practices which confronted us whichever way we turned; instructed in the ways and extent to which these nations for centuries have been and are conserving and utilizing their natural resources, surprised at the magnitude of the returns they are getting from their fields, and amazed at the amount of efficient human labor cheerfully given for a daily wage of five cents and their food, or for fifteen cents, United States currency, without food.

The three main islands of Japan in 1907 had a population of 46,977,003 maintained on 20,000 square miles of cultivated field. This is at the rate of more than three people to each acre, and of 2,349 to each square mile; and yet the total agricultural imports into Japan in 1907 exceeded the agricultural exports by less than one dollar per capita. If the cultivated land of Holland is estimated at but one-third of her total area, the density of her population in 1905 was, on this basis, less than one-third that of Japan in her three main islands. At the same time Japan is feeding 69 horses and 56 cattle, nearly all laboring animals, to each square mile of cultivated field, while we were feeding in 1900 but 30 horses and mules per same area, these being our laboring animals.

As coarse food transformers Japan was maintaining 16,500,000 domestic fowl, 825 per square mile, but only one for almost three of her people. We were maintaining, in 1900, 250,600,000 poultry, but only 387 per square mile of cultivated field and yet more than three for each person. Japan's coarse food transformers in the form of swine, goats and sheep aggregated but 13 to the square mile and provided but one of these units for each 180 of her people; while in the United States in 1900 there were being maintained, as transformers of grass and coarse grain into meat and milk, 95 cattle, 99 sheep and 72 swine per each square mile of improved farms. In this reckoning each of the cattle should be counted as the equivalent of perhaps five of the sheep and swine, for the transforming power of the dairy cow is high. On this basis we are maintaining at the rate of more than 646 of the Japanese units per square mile, and more than five of these to every man, woman and child, instead of one to every 180 of the population, as is the case in Japan.

Correspondingly accurate statistics are not accessible for China but in the Shantung province we talked with a farmer having 12 in his family and who kept one donkey, one cow, both exclusively laboring animals, and two pigs on 2.5 acres of cultivated land where he grew wheat, millet, sweet potatoes and beans. Here is a density of population equal to 3,072 people, 256 donkeys, 256 cattle and 512 swine per square mile. In another instance where the holding was one and two-thirds acres the farmer had 10 in his family and was maintaining one donkey and one pig, giving to this farm land a maintenance capacity of 3,840 people, 384 donkeys and 384 pigs to the square mile, or 240 people, 24 donkeys and 24 pigs to one of our forty-acre farms which our farmers regard too small for a single family. The average of seven Chinese holdings which we visited and where we obtained similar data indicates a maintenance capacity for those lands of 1,783 people, 212 cattle or donkeys and 399 swine,—1,995 consumers and 399 rough food transformers per square mile of farm land. These statements for China represent strictly rural populations. The rural population of the United States in 1900 was placed at the rate of 61 per square mile of improved farm land and there were 30 horses and mules. In Japan the rural population had a density in 1907 of 1,922 per square mile, and of horses and cattle together 125.

The population of the large island of Chungming in the mouth of the Yangtse river, having an area of 270 square miles, possessed, according to the official census of 1902, a density of 3,700 per square mile and yet there was but one large city on the island, hence the population is largely rural.

It could not be other than a matter of the highest industrial, educational and social importance to all nations if there might be brought to them a full and accurate account of all those conditions which have made it possible for such dense populations to be maintained so largely upon the products of Chinese, Korean and Japanese soils. Many of the steps, phases and practices through which this evolution has passed are irrevocably buried in the past but such remarkable maintenance efficiency attained centuries ago and projected into the present with little apparent decadence merits the most profound study and the time is fully ripe when it should be made. Living as we are in the morning of a century of transition from isolated to cosmopolitan national life when profound readjustments. industrial, educational and social, must result, such an investigation cannot be made too soon. It is high time for each nation to study the others and by mutual agreement and co-operative effort, the results of such studies should become available to all concerned, made so in the spirit that each should become coordinate and mutually helpful component factors in the world's progress.

One very appropriate and immensely helpful means for attacking this problem, and which should prove mutually helpful to citizen and state, would be for the higher educational institutions of all nations, instead of exchanging courtesies through their baseball teams, to send select bodies of their best students under competent leadership and by international agreement, both east and west, organizing therefrom investigating bodies each containing components of the eastern and western civilization and whose purpose it should be to study specifically set problems. Such a movement well conceived and directed, manned by the most capable young men, should create an international acquaintance and spread broadcast a body of important knowledge which would develop as the young men mature and contribute immensely toward world peace and world progress. If some broad plan of international effort such as is here suggested were organized the expense of maintenance might well be met by diverting so much as is needful from the large sums set aside for the expansion of navies, for such steps as these, taken in the interests of world uplift and world peace, could not fail to be more efficacious and less expensive than increase in fighting equipment. It would cultivate the spirit of pulling together and of a square deal rather than one of holding aloof and of striving to gain unneighborly advantage.

Many factors and conditions conspire to give to the farms and farmers of the Far East their high maintenance efficiency and some of these may be succinctly stated. The portions of China, Korea and Japan where dense populations have developed and are being maintained occupy exceptionally favorable geographic positions so far as these influence agricultural production. Canton in the south of China has the latitude of Havana, Cuba, while Mukden in Manchuria, and northern Honshu in Japan are only as far north as New York city, Chicago and northen California. The United States lies mainly between 50 degrees and 30 degrees of latitude while these three countries lie between 40 degrees and 20 degrees, some seven hundred miles further south. This difference of position, giving them longer seasons, has made it possible for them to devise systems of agriculture whereby they grow two, three and even four crops on the same piece of ground each year. In southern China, in Formosa and in parts of Japan two crops of rice are grown; in the Chekiang province there may be a crop of rape, of wheat or barley or of windsor beans or clover which is followed in midsummer by another of cotton or of rice. In the Shantung province wheat or barley in the winter and spring may be followed in summer by large or small millet, sweet potatoes, soy beans or pea-nuts. At Tientsin, 39° north, in the latitude of Cincinnati, Indianapolis, and Springfield, Illinois, we talked with a farmer who followed his crop of wheat on his small holding with one of onions and the onions with cabbage, realizing from the three crops at the rate of $163, gold, per acre; and with another who planted Irish potatoes at the earliest opportunity in the spring, marketing them when small, and following these with radishes, the radishes with cabbage, realizing from the three crops at the rate of $203 per acre.

Nearly 500,000,000 people are being maintained, chiefly upon the products of an area smaller than the improved farm lands of the United States. Complete a square on the lines drawn from Chicago southward to the Gulf and west-ward across Kansas, and there will be enclosed an area greater than the cultivated fields of China, Korea and Japan and from which five times our present population are fed.

The rainfall in these countries is not only larger than that even in our Atlantic and Gulf states, but it falls more exclusively during the summer season when its efficiency in crop production may be highest. South China has a rainfall of some 80 inches with little of it during the win-ter, while in our southern states the rainfall is nearer 60 inches with less than one-half of it between June and Sep-tember. Along a line drawn from Lake Superior through central Texas the yearly precipitation is about 30 inches but only 16 inches of this falls during the months May to September; while in the Shantung province, China, with an annual rainfall of little more than 24 inches, 17 of these fall during the months designated and most of this in July and August. When it is stated that under the best tillage and with no loss of water through percolation, most of our agricultural crops require 300 to 600 tons of water for each ton of dry substance brought to maturity, it can be readily understood that the right amount of available moisture, coming at the proper time, must be one of the prime factors of a high maintenance capacity for any soil, and hence that in the Far East, with their intensive methods, it is possible to make their soils yield large returns.

The selection of rice and of the millets as the great staple food crops of these three nations, and the systems of agriculture they have evolved to realize the most from them, are to us remarkable and indicate a grasp of essentials and principles which may well cause western nations to pause and reflect.

Notwithstanding the large and favorable rainfall of these countries, each of the nations have selected the one crop which permits them to utilize not only practically the entire amount of rain which falls upon their fields, but in addition enormous volumes of the run-off from adjacent uncultivable mountain country. Wherever paddy fields are practicable there rice is grown. In the three main islands of Japan 56 per cent of the cultivated fields, 11,000 square miles, is laid out for rice growing and is maintained under water from transplanting to near harvest time, after which the land is allowed to dry, to be devoted to dry land crops during the balance of the year, where the season permits.

To anyone who studies the agricultural methods of the Far East in the field it is evident that these people, centuries ago, came to appreciate the value of water in crop production as no other nations have. They have adapted conditions to crops and crops to conditions until with rice they have a cereal which permits the most intense fertilization and at the same time the ensuring of maximum yields against both drought and flood. With the practice of western nations in all humid climates, no matter how completely and highly we fertilize, in more years than not yields are reduced by a deficiency or an excess of water.

It is difficult to convey, by word or map, an adequate conception of the magnitude of the systems of canalization which contribute primarily to rice culture. A conservative estimate would place the miles of canals in China at fully 200,000 and there are probably more miles of canal in China, Korea and Japan than there are miles of railroad in the United States. China alone has as many acres in rice each year as the United States has in wheat and her annual product is more than double and probably threefold our annual wheat crop, and yet the whole of the rice area produces at least one and sometimes two other crops each year.

The selection of the quick-maturing, drought-resisting millets as the great staple food crops to be grown wherever water is not available for irrigation, and the almost universal planting in hills or drills, permitting intertillage, thus adopting centuries ago the utilization of earth mulches in conserving soil moisture, has enabled these people to secure maximum returns in seasons of drought and where the rainfall is small. The millets thrive in the hot summer climates; they survive when the available soil moisture is reduced to a low limit, and they grow vigorously when the heavy rains come. Thus we find in the Far East, with more rainfall and a better distribution of it than occurs in the United States, and with warmer, longer seasons, that these people have with rare wisdom combined both irrigation and dry farming methods to an extent and with an intensity far beyond anything our people have ever dreamed, in order that they might maintain their dense populations.

Notwithstanding the fact that in each of these countries the soils are naturally more than ordinarily deep, inherently fertile and enduring, judicious and rational methods of fertilization are everywhere practiced; but not until recent years, and only in Japan, have mineral commercial fertilizers been used. For centuries, however, all cultivated lands, including adjacent hill and mountain sides, the canals, streams and the sea have been made to contribute what they could toward the fertilization of cultivated fields and these contributions in the aggregate have been large. In China, in Korea and in Japan all but the inaccessible portions of their vast extent of mountain and hill lands have long been taxed to their full capacity for fuel, lumber and herbage for green manure and compost material; and the ash of practically all of the fuel and of all of the lumber used at home finds its way ultimately to the fields as fertilizer.

In China enormous quantities of canal mud are applied to the fields, sometimes at the rate of even 70 and more tons per acre. So, too, where there are no canals, both soil and subsoil are carried into the villages and there between the intervals when needed they are, at the expense of great labor, composted with organic refuse and often afterwards dried and pulverized before being carried back and used on the fields as home-made fertilizers. Manure of all kinds, human and animal, is religiously saved and applied to the fields in a manner which secures an efficiency far above our own practices. Statistics obtained through the Bureau of Agriculture, Japan, place the amount of human waste in that country in 1908 at 23,950,295 tons, or 1.75 tons per acre of her cultivated land. The International Concession of the city of Shanghai, in 1908, sold to a Chinese contractor the privilege of entering residences and public places early in the morning of each day in the year and removing the night soil, receiving therefor more than $31,000, gold, for 78,000 tons of waste. All of this we not only throw away but expend much larger sums in doing so.

Japan's production of fertilizing material, regularly prepared and applied to the land annually, amounts to more than 4.5 tons per acre of cultivated field exclusive of

the commercial fertilizers purchased. Between Shanhaikwan and Mukden in Manchuria we passed, on June 18th,
thousands of tons of the dry highly nitrified compost soil
recently carried into the fields and laid down in piles where
it was waiting to be "fed to the crops."

It was not until 1888, and then after a prolonged war
of more than thirty years, generaled by the best scientists
of all Europe, that it was finally conceded as demonstrated
that leguminous plants acting as hosts for lower organisms
living on their roots are largely responsible for the maintenance of soil nitrogen, drawing it directly from the air to
which it is returned through the processes of decay. But
centuries of practice had taught the Far East farmers that
the culture and use of these crops are essential to enduring
fertility, and so in each of the three countries the growing
of legumes in rotation with other crops very extensively
for the express purpose of fertilizing the soil is one of their
old, fixed practices.

Just before, or immediately after the rice crop is harvested, fields are often sowed to "clover" (*Astragalus
sinicus*) which is allowed to grow until near the next transplanting time when it is either turned under directly, or
more often stacked along the canals and saturated while
doing so with soft mud dipped from the bottom of the
canal. After fermenting twenty or thirty days it is applied to the field. And so it is literally true that these old
world farmers whom we regard as ignorant, perhaps because they do not ride sulky plows as we do, have long included legumes in their crop rotation, regarding them as
indispensable.

Time is a function of every life process as it is of every
physical, chemical and mental reaction. The husbandman
is an industrial biologist and as such is compelled to shape
his operations so as to conform with the time requirements
of his crops. The oriental farmer is a time economizer beyond all others. He utilizes the first and last minute and
all that are between. The foreigner accuses the Chinaman
of being always long on time, never in a fret, never in a

hurry. This is quite true and made possible for the reason that they are a people who definitely set their faces toward the future and lead time by the forelock. They have long realized that much time is required to transform organic matter into forms available for plant food and although they are the heaviest users in the world, the largest portion of this organic matter is predigested with soil or subsoil before it is applied to their fields, and at an enormous cost of human time and labor, but it practically lengthens their growing season and enables them to adopt a system of multiple cropping which would not otherwise be possible. By planting in hills and rows with intertillage it is very common to see three crops growing upon the same field at one time, but in different stages of maturity, one nearly ready to harvest; one just coming up, and the other at the stage when it is drawing most heavily upon the soil. By such practice, with heavy fertilization, and by supplemental irrigation when needful, the soil is made to do full duty throughout the growing season.

Then, notwithstanding the enormous acreage of rice planted each year in these countries, it is all set in hills and every spear is transplanted. Doing this, they save in many ways except in the matter of human labor, which is the one thing they have in excess. By thoroughly preparing the seed bed, fertilizing highly and giving the most careful attention, they are able to grow on one acre, during 30 to 50 days, enough plants to occupy ten acres and in the mean time on the other nine acres crops are maturing, being harvested and the fields being fitted to receive the rice when it is ready for transplanting, and in effect this interval of time is added to their growing season.

Silk culture is a great and, in some ways, one of the most remarkable industries of the Orient. Remarkable for its magnitude; for having had its birthplace apparently in oldest China at least 2700 years B. C.; for having been laid on the domestication of a wild insect of the woods; and for having lived through more than 4000 years, expanding until a million-dollar cargo of the product has been laid down on our western coast and rushed by special fast express to the east for the Christmas trade.

A low estimate of China's production of raw silk would be 120,000,000 pounds annually, and this with the output of Japan, Korea and a small area of southern Manchuria, would probably exceed 150,000,000 pounds annually, representing a total value of perhaps $700,000,000, quite equalling in value the wheat crop of the United States, but produced on less than one-eighth the area of our wheat fields.

The cultivation of tea in China and Japan is another of the great industries of these nations, taking rank with that of sericulture if not above it in the important part it plays in the welfare of the people. There is little reason to doubt that this industry has its foundation in the need of something to render boiled water palatable for drinking purposes. The drinking of boiled water is universally adopted in these countries as an individually available and thoroughly efficient safeguard against that class of deadly disease germs which thus far it has been impossible to exclude from the drinking water of any densely peopled country.

Judged by the success of the most thorough sanitary measures thus far instituted, and taking into consideration the inherent difficulties which must increase enormously with increasing populations, it appears inevitable that modern methods must ultimately fail in sanitary efficiency and that absolute safety can be secured only in some manner having the equivalent effect of boiling drinking water, long ago adopted by the Mongolian races.

In the year 1907 Japan had 124,482 acres of land in tea plantations, producing 60,877,975 pounds of cured tea. In China the volume annually produced is much larger than that of Japan, 40,000,000 pounds going annually to Tibet alone from the Szechwan province; and the direct export to foreign countries was, in 1905, 176,027,255 pounds, and in 1906 it was 180,271,000, so that their annual export must exceed 200,000,000 pounds with a total annual output more than double this amount of cured tea.

But above any other factor, and perhaps greater than all of them combined in contributing to the high maintenance efficiency attained in these countries must be placed the standard of living to which the industrial classes have been compelled to adjust themselves, combined with their remarkable industry and with the most intense economy they practice along every line of effort and of living.

Almost every foot of land is made to contribute material for food, fuel or fabric. Everything which can be made edible serves as food for man or domestic animals. Whatever cannot be eaten or worn is used for fuel. The wastes of the body, of fuel and of fabric worn beyond other use are taken back to the field; before doing so they are housed against waste from weather, compounded with intelligence and forethought and patiently labored with through one, three or even six months, to bring them into the most efficient form to serve as manure for the soil or as feed for the crop. It seems to be a golden rule with these industrial classes, or if not golden, then an inviolable one, that whenever an extra hour or day of labor can promise even a little larger return then that shall be given, and neither a rainy day nor the hottest sunshine shall be permitted to cancel the obligation or defer its execution.

I.

FIRST GLIMPSES OF JAPAN.

We left the United States from Seattle for Shanghai, China, sailing by the northern route, at one P. M. February second, reaching Yokohama February 19th and Shanghai, March 1st. It was our aim throughout the journey to keep in close contact with the field and crop problems and to converse personally, through interpreters or otherwise, with the farmers, gardeners and fruit growers themselves; and we have taken pains in many cases to visit the same fields or the same region two, three or more times at different intervals during the season in order to observe different phases of the same cultural or fertilization methods as these changed or varied with the season.

Our first near view of Japan came in the early morning of February 19th when passing some three miles off the point where the Pacific passenger steamer Dakota was beached and wrecked in broad daylight without loss of life two years ago. The high rounded hills were clothed neither in the dense dark forest green of Washington and Vancouver, left sixteen days before, nor yet in the brilliant emerald such as Ireland's hills in June fling in unparalleled greeting to passengers surfeited with the dull grey of the rolling ocean. This lack of strong forest growth and even of shrubs and heavy herbage on hills covered with deep soil, neither cultivated nor suffering from serious erosion, yet surrounded by favorable climatic conditions, was our first great surprise.

To the southward around the point, after turning north-ward into the deep bay, similar conditions prevailed, and at ten o'clock we stood off Uraga where Commodore Perry anchored on July 8th, 1853, bearing to the Shogun President Fillmore's letter which opened the doors of Japan to the commerce of the world and, it is to be hoped brought to her people, with their habits of frugality and industry so indelibly fixed by centuries of inheritance, better opportunities for development along those higher lines destined to make life still more worth living.

As the Tosa Maru drew alongside the pier at Yokohama it was raining hard and this had attired an army after the manner of Robinson Crusoe, dressed as seen in Fig. 1, ready to carry you and yours to the Customs house and beyond for one, two, three or five cents. Strong was the contrast when the journey was reversed and we descended the gang plank at Seattle, where no one sought the opportunity of moving baggage.

Through the kindness of Captain Harrison of the Tosa Maru in calling an interpreter by wireless to meet the steamer, it was possible to utilize the entire interval of stop in Yokohama to the best advantage in the fields and gardens spread over the eighteen miles of plain extending to Tokyo, traversed by both electric tram and railway lines, each running many trains making frequent stops; so that this wonderfully fertile and highly tilled district could be readily and easily reached at almost any point.

We had left home in a memorable storm of snow, sleet and rain which cut out of service telegraph and telephone lines over a large part of the United States; we had sighted the Aleutian Islands, seeing and feeling nothing on the way which could suggest a warm soil and green fields, hence our surprise was great to find the jinricksha men with bare feet and legs naked to the thighs, and greater still when we found, before we were outside the city limits, that the electric tram was running between fields and gardens green with wheat, barley, onions, carrots, cabbage and other vegetables. We were rushing through the Orient

Fig. 1.—Rainy weather costume, as worn in Japan and typical of those used under similar conditions in both Korea and China. The picture shows a group of Japanese rice field laborers with their most common tools.

with everything outside the car so strange and different from home that the shock came like a bolt of lightning out of a clear sky.

In the car every man except myself and one other was smoking tobacco and that other was inhaling camphor through an ivory mouthpiece resembling a cigar holder closed at the end. Several women, tiring of sitting foreign style, slipped off—I cannot say out of—their shoes and sat facing the windows, with toes crossed behind them on the seat. The streets were muddy from the rain and everybody Japanese was on rainy-day wooden shoes, the soles carried three to four inches above the ground by two cross blocks, in the manner seen in Fig. 2. A mother, with baby on her back and a daughter of sixteen years came into the car. Notwithstanding her high shoes the mother had dipped one toe into the mud. Seated, she slipped her foot off. Without evident instructions the pretty black-eyed, glossy-haired, red-lipped lass, with cheeks made rosy, picked up the shoe, withdrew a piece of white tissue paper from the great pocket in her sleeve, deftly cleaned the otherwise spotless white cloth sock and then the shoe, threw the paper on the floor, looked to see that her fingers were not soiled, then set the shoe at her mother's foot, which found its place without effort or glance.

Everything here was strange and the scenes shifted with the speed of the wildest dream. Now it was driving piles for the foundation of a bridge. A tripod of poles was erected above the pile and from it hung a pulley. Over the pulley passed a rope from the driving weight and from its end at the pulley ten cords extended to the ground. In a circle at the foot of the tripod stood ten agile Japanese women. They were the hoisting engine. They chanted in perfect rhythm, hauled and stepped, dropped the weight and hoisted again, making up for heavier hammer and higher drop by more blows per minute. When we reached Shanghai we saw the pile driver being worked from above. Fourteen Chinese men stood upon a raised staging, each with a separate cord passing direct from the hand to the

2

weight below. A concerted, half-musical chant, modulated
to relieve monotony, kept all hands together. What did
the operation of this machine cost? Thirteen cents, gold,

Fig. 2.—Girl on rainy-day wooden shoes, carrying and entertaining child
in the way most common in Japan.

per man per day, which covered fuel and lubricant, both
automatically served. Two additional men managed the
piles, two directed the hammer, eighteen manned the outfit. Two dollars and thirty-four cents per day covered
fuel, superintendence and repairs. There was almost no

capital invested in machinery. Men were plenty and to spare. Rice was the fuel, cooked without salt, boiled stiff, reenforced with a bit of pork or fish, appetized with salted cabbage or turnip and perhaps two or three of forty and more other vegetable relishes. And are these men strong and happy? They certainly were strong. They are steadily increasing their millions, and as one stood and watched them at their work their faces were often wreathed in smiles and wore what seemed a look of satisfaction and contentment.

Among the most common sights on our rides from Yokohama to Tokyo, both within the city and along the roads leading to the fields, starting early in the morning, were the loads of night soil carried on the shoulders of men and on the backs of animals, but most commonly on strong carts drawn by men, bearing six to ten tightly covered wooden containers holding forty, sixty or more pounds each. Strange as it may seem, there are not today and apparently never have been, even in the largest and oldest cities of Japan, China or Korea, anything corresponding to the hydraulic systems of sewage disposal used now by western nations. Provision is made for the removal of storm waters but when I asked my interpreter if it was not the custom of the city during the winter months to discharge its night soil into the sea, as a quicker and cheaper mode of disposal, his reply came quick and sharp, "No, that would be waste. We throw nothing away. It is worth too much money." In such public places as railway stations provision is made for saving, not for wasting, and even along the country roads screens invite the traveler to stop, primarily for profit to the owner, more than for personal convenience.

Between Yokohama and Tokyo, along the electric car line and not far distant from the seashore, there were to be seen in February very many long, fence-high screens extending east and west, strongly inclined to the north, and built out of rice straw,closely tied together and supported on bamboo poles carried upon posts of wood set in the ground. These

Fig. 3.—Method of drying seaweed used for food. The small black squares on the larger light ones are the seaweed. The skewers seen pin the squares of matting against the long screens, six of which are shown in parallel series.

Fig. 4.—Section of shallow sea bottom planted to brushwood on which the edible seaweeds attach themselves and grow.

screens, set in parallel series of five to ten or more in number and several hundred feet long, were used for the purpose of drying varieties of delicate seaweed, these being spread out in the manner shown in Fig. 3.

The seaweed is first spread upon separate ten by twelve inch straw mats, forming a thin layer seven by eight inches. These mats are held by means of wooden skewers forced through the body of the screen, exposing the seaweed to the direct sunshine. After becoming dry the rectangles of seaweed are piled in bundles an inch thick, cut once in two, forming packages four by seven inches, which are neatly tied and thus exposed for sale as soup stock and for other purposes.

To obtain this seaweed from the ocean small shrubs and the limbs of trees are set up in the bottom of shallow water, as seen in Fig. 4. To these limbs the seaweeds become attached, grow to maturity and are then gathered by hand. By this method of culture large amounts of important food stuff are grown for the support of the people on areas otherwise wholly unproductive.

Another rural feature, best shown by photograph taken in February, is the method of training pear orchards in Japan, with their limbs tied down upon horizontal overhead trellises at a hight under which a man can readily walk erect and easily reach the fruit with the hand while standing upon the ground. Pear orchards thus form arbors of greater or less size, the trees being set in quincunx order about twelve feet apart in and between the rows. Bamboo poles are used overhead and these carried on posts of the same material 1.5 to 2.5 inches in diameter, to which they are tied. Such a pear orchard is shown in Fig. 5.

The limbs of the pear trees are trained strictly in one plane, tying them down and pruning out those not desired. As a result the ground beneath is completely shaded and every pear is within reach, which is a great convenience when it becomes desirable to protect the fruit

Fig. 5.—Looking down upon an extensive pear orchard whose limbs are trained horizontally, forming an arbor completely shading the ground when in leaf, and placing all of the fruit within reach of the hand from beneath.

Fig. 6.—Pear trees at Akashi Experiment Station, Japan. Pears protected by paper bags. Special form of pruning advised by Prof. Ono, standing on the left, with Prof. Tokito. The trees branch below rather than at the level of the trellis.

from insects, by tying paper bags over every pear as seen in Figs. 6 and 7. The orchard ground is kept free from weeds and not infrequently is covered with a layer of rice or other straw, extensively used in Japan as a ground cover with various crops and when so used is carefully laid in handfuls from bundles, the straws being kept parallel as when harvested.

Fig. 7.—Low branching pear orchard with pears protected by paper bags, at Akashi Experiment Station, Japan.

To one from a country of 160-acre farms, with roads four rods wide; of cities with broad streets and residences with green lawns and ample back yards; and where the cemeteries are large and beautiful parks, the first days of travel in these old countries force the over-crowding upon the attention as nothing else can. One feels that the cities are greatly over-crowded with houses and shops, and these with people and wares; that the country is over-crowded with fields and the fields with crops; and that in Japan the over-crowding is greatest of all in the cemeteries,

Fig. 8.—Street in a Hakone country village. The general abs ence of old forest growth on the hills recently cut over is characteristic of much of Japan.

gravestones almost touching and markers for families literally in bundles at a grave, while round about there may be no free country whatever, dwellings, gardens or rice paddies contesting the tiny allotted areas too closely to leave even foot-paths between.

Unless recently modified through foreign influence the streets of villages and cities are narrow, as seen in Fig. 8, where however the street is unusually broad. This is a village in the Hakone district on a beautiful lake of the same name, where stands an Imperial summer palace, seen near the center of the view on a hill across the lake. The roofs of the houses here are typical of the neat, careful thatching with rice straw, very generally adopted in place of tile for the country villages throughout much of Japan. The shops and stores, open full width directly upon the street, are filled to overflowing, as seen in Fig. 9 and in Fig. 22.

Fig. 9.—Small store full to overflowing; entire front opening flush with the street.

Fig. 10.—Chinese country village lining both sides of a canal. Section one-third of a mile long between two bridges, where in three rows of houses live 240 families.

In the canalized regions of China the country villages crowd both banks of a canal, as is the case in Fig. 10. Here, too, often is a single street and it very narrow, very crowded and very busy. Stone steps lead from the houses down into the water where clothing, vegetables, rice and what not are conveniently washed. In this particular village two rows of houses stand on one side of the canal separated by a very narrow street, and a single row on the other. Between the bridge where the camera was exposed and one barely discernible in the background, crossing the canal a third of a mile distant, we counted upon one side, walking along the narrow street, eighty houses each with its family, usually of three generations and often of four. Thus in the narrow strip, 154 feet broad, including 16 feet of street and 30 feet of canal, with its three lines of houses, lived no less than 240 families and more than 1200 and probably nearer 2000 people.

When we turn to the crowding of fields in the country nothing except seeing can tell so forcibly the fact as such landscapes as those of Figs. 11, 12 and 13, one in Japan, one in Korea and one in China, not far from Nanking, looking from the hills across the fields to the broad Yangtse kiang, barely discernible as a band of light along the horizon.

The average area of the rice field in Japan is less than five square rods and that of her upland fields only about twenty. In the case of the rice fields the small size is necessitated partly by the requirement of holding water on the sloping sides of the valley, as seen in Fig. 11. These small areas do not represent the amount of land worked by one family, the average for Japan being more nearly 2.5 acres. But the lands worked by one family are seldom contiguous, they may even be widely scattered and very often rented.

The people generally live in villages, going often considerable distances to their work. Recognizing the great disadvantage of scattered holdings broken into such small areas, the Japanese Government has passed laws for the adjust-

Fig. 11.—Closely crowded fields of rice in Japan, each rice paddy filled with water and recently transplanted.

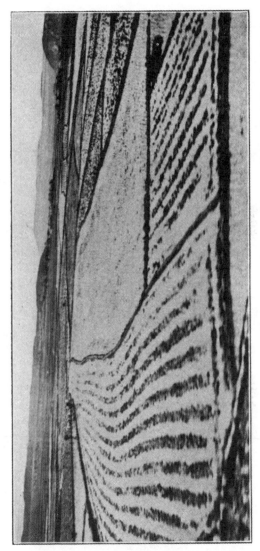

Fig. 12.—Landscape in Korea, showing subdivision of the valley surface into small irregular fields separated only by narrow, low ridges of earth scarcely more than a foot wide and high. The center field is planted to rice, fields on the right are plowed and watered but not fitted, the ridged field on the left is watered but not plowed.

Fig. 13.—Landscape of rice fields in China. Fields in the foreground still covered with winter crops, but when harvested, to be planted to rice. White areas flooded with water and fitting for rice. Yangtse river near horizon.

ment of farm lands which have been in force since 1900. It provides for the exchange of lands; for changing boundaries; for changing or abolishing roads, embankments, ridges or canals and for alterations in irrigation and drainage which would ensure larger areas with channels and roads straightened, made less numerous and less wasteful of time, labor and land. Up to 1907 Japan had issued permits for the readjustment of over 240,000 acres, and Fig. 14 is a landscape in one of these readjusted districts. To provide capable experts for planning and supervising these changes the Government in 1905 intrusted the training of men to the higher agricultural school belonging to the Dai Nippon Agricultural Association and since 1906 the Agricultural College and the Kogyokusha have undertaken the same task and now there are men sufficient to push the work as rapidly as desired.

It may be remembered, too, as showing how, along other fundamental lines, Japan is taking effective steps to improve the condition of her people, that she already has her Imperial highways extending from one province to another; her prefectural roads which connect the cities and villages within the prefecture; and those more local which serve the farms and villages. Each of the three systems of roads is maintained by a specific tax levied for the purpose which is expended under proper supervision, a designated section of road being kept in repair through the year by a specially appointed crew, as is the practice in railroad maintenance. The result is, Japan has roads maintained in excellent condition, always narrow, sacrificing the minimum of land, and everywhere without fences.

How the fields are crowded with crops and all available land is made to do full duty in these old, long-tilled countries is evident in Fig. 15 where even the narrow dividing ridges but a foot wide, which retain the water on the rice paddies, are bearing a heavy crop of soy beans; and where may be seen the narrow pear orchard standing on the very slightest rise of ground, not a foot above the water all around, which could better be left in grading the paddies to proper level.

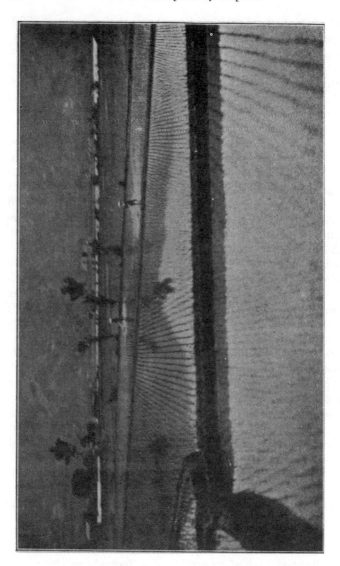

Fig. 14.—Landscape in one of the readjusted districts in Japan where division lines between paddy fields have been straightened. Men using new rice-weeding cultivators.

Fig. 15.—The entire field completely occupied by crops, rendering effective service. Soy beans on the dividing lines, rice in the paddies, pear orchard on the narrow raised ridge.

3

Fig. 16.—Young peach orchard doing intense duty as a market garden, growing peas, cabbage and windsor beans.

How closely the ground itself may be crowded with plants is seen in Fig. 16, where a young peach orchard, whose tree tops were six feet through, planted in rows twenty-two feet apart, had also ten rows of cabbage, two rows of large windsor beans and a row of garden peas. Thirteen rows of vegetables in 22 feet, all luxuriant and strong, and note the judgment shown in placing the tallest plants, needing the most sun, in the center between the trees.

But these old people, used to crowding and to being crowded, and long ago capable of making four blades of grass grow where Nature grew but one, have also learned how to double the acreage where a crop needs more elbow than it does standing room, as seen in Fig. 17. This man's garden had an area of but 63 by 68 feet and two square rods of this was held sacred to the family grave mound, and yet his statement of yields, number of crops and prices made his earning $100 a year on less than one-tenth of an acre.

His crop of cucumbers on less than .06 of an acre would bring him $20. He had already sold $5 worth of greens and a second crop would follow the cucumbers. He had just irrigated his garden from an adjoining canal, using a foot-power pump, and stated that until it rained he would repeat the watering once per week. It was his wife who stood in the garden and, although wearing trousers, her dress showed full regard for modesty.

But crowding crops more closely in the field not only requires higher feeding to bring greater returns, but also relatively greater care, closer watchfulness in a hundred ways and a patience far beyond American measure; and so, before the crowding of the crops in the field and along with it, there came to these very old farmers a crowding of the grey matter in the brain with the evolution of effective texture. This is shown in his fields which crowd the landscape. It is seen in the crops which crowd his fields. You see it in the old man's face, Fig. 18, standing opposite his compeer, Prince Ching, Fig. 19, each clad in

Fig. 17.—Increasing the available surface of the field so that double the number of plants may occupy the ground. A row of cucumbers on opposite sides of each trellis will cover its surface.

winter dress which is the embodiment of conversation, retaining the fires of the body for its own needs, to release the growth on mountain sides for other uses. And when one realizes how, nearly to the extreme limits, conservation along all important lines is being practiced as an inherited instinct, there need be no surprise when one reflects that the two men, one as feeder and the other as leader, are standing in the fore of a body of four hundred millions of people who have marched as a nation through perhaps forty centuries, and who now, in the light and great promise of unfolding science have their faces set toward a still more hopeful and longer future.

On February 21st the Tosa Maru left Yokohama for Kobe at schedule time on the tick of the watch, as she had done from Seattle. All Japanese steamers appear to be moved with the promptness of a railway train. On reaching Kobe we transferred to the Yamaguchi Maru which sailed the following morning, to shorten the time of reaching Shanghai. This left but an afternoon for a trip into the country between Kobe and Osaka, where we found, if possible, even higher and more intensive culture practices than on the Tokyo plain, there being less land not carrying a winter crop. And Fig. 20 shows how closely the crops crowd the houses and shops. Here were very many cement lined cisterns or sheltered reservoirs for collecting manures and preparing fertilizers and the appearance of both soil and crops showed in a marked manner to what advantage. We passed a garden of nearly an acre entirely devoted to English violets just coming into full bloom. They were grown in long parallel east and west beds about three feet wide. On the north edge of each bed was erected a rice-straw screen four feet high which inclined to the south, overhanging the bed at an angle of some thirty-five degrees, thus forming a sort of bake-oven tent which reflected the sun, broke the force of the wind and checked the loss of heat absorbed by the soil.

The voyage from Kobe to Moji was made between 10 in the morning, February 24th, and 5:30 P. M. of February

Fig. 18.—Aged Chinese farmer in winter dress, who leads
in the maintenance of his nation.

Fig. 19.—Prince Ching, also in winter dress.

25th over a quiet sea with an enjoyable ride. Being fog-
bound during the night gave us the whole of Japan's
beautiful Inland Sea, enchanting beyond measure, in all
its near and distant beauty but which no pen, no brush,
no camera may attempt. Only the eye can convey. Before
reaching harbor the tide had been rising and the strait
separating Honshu from Kyushu island was running like
a mighty swirling river between Moji and Shimonoseki,
dangerous to attempt in the dark, so we waited until
morning.

There was cargo to take on board and the steamer must
coal. No sooner had the anchor dropped and the steamer
swung into the current than lighters came alongside with
out-going freight. The small, strong, agile Japanese stevedores had this task completed by 8:30 P. M. and when
we returned to the deck after supper another scene was
on. The cargo lighters had gone and four large barges
bearing 250 tons of coal had taken their places on opposite
sides of the steamer, each illuminated with buckets of
blazing coal or by burning conical heaps on the surface.
From the bottom of these pits in the darkness the illumina-
tion suggested huge decapitated ant heaps in the wildest
frenzy, for the coal seemed covered and there was hurry
in every direction. Men and women, boys and girls, bending to their tasks, were filling shallow saucer-shaped baskets
with coal and stacking them eight to ten high in a semicircle, like coin for delivery. Rising out of these pits
sixteen feet up the side of the steamer and along her deck
to the chutes leading to her bunkers were what seemed
four endless human chains, in service the prototype of our
modern conveyors, but here each link animated by its own
power. Up these conveyors the loaded buckets passed, one
following another at the rate of 40 to 60 per minute, to
return empty by the descending line, and over the four
chains one hundred tons per hour, for 250 tons of coal
passed to the bunkers in two and a half hours. Both men
and women stood in the line and at the upper turn of one
of these, emptying the buckets down the chute, was a

Fig. 20.—Newly started gardens crowded close about houses and shops between Kobe and Osaka, Japan, Feb. 23.

mother with her two-year-old child in the sling on her
back, where it rocked and swayed to and fro, happy the
entire time. It was often necessary for the mother to
adjust her baby in the sling whenever it was leaning
uncomfortably too far to one side or the other, but she
did it skillfully, always with a shrug of the shoulders,
for both hands were full. The mother looked strong,
was apparently accepting her lot as a matter of course
and often, with a smile, turned her face to the child,
who patted it and played with her ears and hair.
Probably her husband was doing his part in a more
strenuous place in the chain and neither had time to be
troubled with affinities for it was 10:30 P. M. when the
baskets stopped, and somewhere no doubt there was a home
to be reached and perhaps supper to get. Shall we be able,
when our numbers have vastly increased, to permit all
needful earnings to be acquired in a better way?

  We left Moji in the early morning and late in the
evening of the same day entered the beautiful harbor of
Nagasaki, all on board waiting until morning for a launch
to go ashore. We were to sail again at noon so available
time for observation was short and we set out in a ricksha
at once for our first near view of terraced gardening on
the steep hillsides in Japan. In reaching them and in
returning our course led through streets paved with long,
thick and narrow stone blocks, having deep open gutters
on one or both sides close along the houses, into which waste
water was emptied and through which the storm waters
found their way to the sea. Few of these streets were
more than twelve feet wide and close watching, with much
dodging, was required to make way through them. Here,
too, the night soil of the city was being removed in closed
receptacles on the shoulders of men, on the backs of horses
and cattle and on carts drawn by either. Other men and
women were hurrying along with baskets of vegetables well
illustrated in Fig. 21, some with fresh cabbage, others with
high stacks of crisp lettuce, some with monstrous white
radishes or turnips, others with bundles of onions, all com-

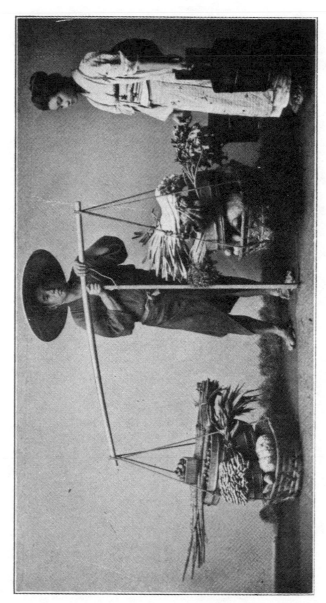

Fig. 21.—Vegetable vender with his load as carried from house to house.

Fig. 22.—Store of Japanese vegetable dealer. The large vegetables in the center are bamboo sprouts, of the nature of asparagus, used extensively for food everywhere in the Orient.

ing down from the terraced gardens to the markets. We passed loads of green bamboo poles just cut, three inches in diameter at the butt and twenty feet long, drawn on carts. Both men and women were carrying young children and older ones were playing and singing in the street. Very many old women, some feeble looking, moved,loaded,through the throng. Homely little dogs, an occasional lean cat, and hens and roosters scurried across the street from one low market or store to another. Back of the rows of small stores and shops fronting on the clean narrow streets were the dwellings whose exits seemed to open through the stores, few or no open courts of any size separating them from the market or shop. The opportunity which the oriental house-wife may have in the choice of vegetables on going to the market, and the attractive manner of displaying such products in Japan, are seen in Fig. 22.

We finally reached one of the terraced hillsides which rise five hundred to a thousand feet above the harbor with sides so steep that garden areas have a width of seldom more than twenty to thirty feet and often less, while the front of each terrace may be a stone wall, sometimes twelve feet high, often more than six, four and five feet being the most common hight. One of these hillside slopes is seen in Fig. 23. These terraced gardens are both short and narrow and most of them bounded by stone walls on three sides, suggesting house foundations, the two end walls sloping down the hill from the hight of the back terrace, dropping to the ground level in front, these forming foot-paths leading up the slope occasionally with one, two or three steps in places.

Each terrace sloped slightly down the hill at a small angle and had a low ridge along the front. Around its entire border a narrow drain or furrow was arranged to collect surface water and direct it to drainage channels or into a catch basin where it might be put back on the garden or be used in preparing liquid fertilizer. At one corner of many of these small terraced gardens were cement lined pits, used both as catch basins for water and as receptacles

Fig. 23.—Terraced gardens on hillside at Nagasaki. Japan.

for liquid manure or as places in which to prepare compost. Far up the steep paths, too, along either side, we saw many piles of stable manure awaiting application, all of which had been brought up the slopes in baskets on bamboo poles, carried on the shoulders of men and women.

II.

GRAVE LANDS OF CHINA.

The launch had returned the passengers to the steamer at 11:30; the captain was on the bridge; prompt to the minute at the call ''Hoist away'' the signal went below and the Yamaguchi's whistle filled the harbor and overflowed the hills. The cable wound in, and at twelve, noon, we were leaving Nagasaki, now a city of 153,000 and the western doorway of a nation of fifty-one millions of people but of little importance before the sixteenth century when it became the chief mart of Portuguese trade. We were to pass the Koreans on our right and enter the portals of a third nation of four hundred millions. We had left a country which had added eighty-five millions to its population in one hundred years and which still has twenty acres for each man, woman and child, to pass through one which has but one and a half acres per capita, and were going to another whose allotment of acres, good and bad, is less than 2.4. We had gone from practices by which three generations had exhausted strong virgin fields, and were coming to others still fertile after thirty centuries of cropping. On January 30th we crossed the head waters of the Mississippi-Missouri, four thousand miles from its mouth, and on March 1st were in the mouth of the Yangtse river whose waters are gathered from a basin in which dwell two hundred millions of people

The Yamaguchi reached Woosung in the night and anchored to await morning and tide before ascending the Hwangpoo, believed by some geographers to be the middle

Fig. 24.—Views of grave lands in the delta region of the Yangtse kiang, China.

4

of three earlier delta arms of the Yangtse kiang, the southern entering the sea at Hangchow 120 miles further south, the third being the present stream. As we wound through this great delta plain toward Shanghai, the city of foreign concessions to all nationalities, the first striking feature was the "graves of the fathers", of "the ancestors". At first the numerous grass-covered hillocks dotting the plain seemed to be stacks of grain or straw; then came the query whether they might not be huge compost heaps awaiting distribution in the fields, but as the river brought us nearer to them we seemed to be moving through a land of ancient mound builders and Fig. 24 shows, in its upper section, their appearance as seen in the distance.

As the journey led on among the fields, so large were the mounds, often ten to twelve feet high and twenty or more feet at the base; so grass-covered and apparently neglected; so numerous and so irregularly scattered, without apparent regard for fields, that when we were told these were graves we could not give credence to the statement, but before the city was reached we saw places where, by the shifting of the channel, the river had cut into some of these mounds, exposing brick vaults, some so low as to be under water part of the time, and we wonder if the fact does not also record a slow subsidence of the delta plain under the ever increasing load of river silt.

A closer view of these graves in the same delta plain is given in the lower section of Fig. 24, where they are seen in the midst of fields and to occupy not only large areas of valuable land but to be much in the way of agricultural operations. A still closer view of other groups, with a farm village in the background, is shown in the middle section of the same illustration, and here it is better seen how large is the space occupied by them. On the right in the same view may be seen a line of six graves surmounting a common lower base which is a type of the larger and higher ones so suggestive of buildings seen in the horizon of the upper section.

Everywhere we went in China, about all of the very old and large cities, the proportion of grave land to cultivated fields is very large. In the vicinity of Canton Christian college, on Honam island, more than fifty per cent of the land was given over to graves and in many places they were so close that one could step from one to another. They are on the higher and dryer lands, the cultivated areas occupying ravines and the lower levels to which water may be more easily applied and which are the most productive. Hilly lands not so readily cultivated, and especially if within reach of cities, are largely so used, as seen in Fig. 25, where the graves are marked by excavated shelves rather than by mounds, as on the plains. These grave lands are not altogether unproductive for they are generally

Fig. 25.—Goats pasturing on grave land near Shanghai, and graves in hilly lands near Canton.

overgrown with herbage of one or another kind and used as pastures for geese, sheep, goats and cattle, and it is not at all uncommon, when riding along a canal, to see a huge water buffalo projected against the sky from the summit of one of the largest and highest grave mounds within reach.   If the herbage is not fed off by animals it is usually cut for feed, for fuel, for green manure or for use in the production of compost to enrich the soil.

Caskets may be placed directly upon the surface of a field, encased in brick vaults with tile roofs, forming such clusters as was seen on the bank of the Grand Canal in Chekiang province, represented in the lower section of Fig. 26, or they may stand singly in the midst of a garden, as

Fig. 26.—Cluster of graves in brick vaults, lower section; and isolated grave in garden, with two large grave mounds, upper section.

in the upper section of the same figure; in a rice paddy
entirely surrounded by water parts of the year, and indeed
in almost any unexpected place. In Shanghai in 1898,
2,763 exposed coffined corpses were removed outside the
International Settlement or buried by the authorities.

Further north, in the Shantung province, where the dry
season is more prolonged and where a severe drought had
made grass short, the grave lands had become nearly naked
soil, as seen in Fig. 27 where a Shantung farmer had just
dug a temporary well to irrigate his little field of barley.
Within the range of the camera, as held to take this view,
more than forty grave mounds besides the seven near by,
are near enough to be fixed on the negative and be discernible under a glass, indicating what extensive areas of land,
in the aggregate, are given over to graves.

Still further north, in Chihli, a like story is told in, if
possible, more emphatic manner and fully vouched for in
the next illustration, Fig. 28, which shows a typical family
group, to be observed in so many places between Taku and
Tientsin and beyond toward Peking. As we entered the
mouth of the Pei-ho for Tientsin, far away to the vanishing
horizon there stretched an almost naked plain except for
the vast numbers of these "graves of the fathers", so
strange, so naked, so regular in form and so numerous that
more than an hour of our journey had passed before we
realized that they were graves and that the country here
was perhaps more densely peopled with the dead than with
the living. In so many places there was the huge father
grave, often capped with what in the distance suggested
a chimney, and the many associated smaller ones, that it
was difficult to realize in passing what they were.

It is a common custom, even if the residence has been
permanently changed to some distant province, to take the
bodies back for interment in the family group; and it is
this custom which leads to the practice of choosing a temporary location for the body, waiting for a favorable opportunity to remove it to the family group. This is often the
occasion for the isolated coffin so frequently seen under a

Fig. 27.—Graves surrounded by fields in the Shantung province. The farmer has dug a temporary well to irrigate the little barley field threatened by drought.

Fig. 28.—Family group of grave mounds in Chihli, between Taku and Tientsin; the largest or father grave is in the rear, those of his two sons standing next.

simple thatch of rice straw, as in Fig. 29; and the many small stone jars containing skeletons of the dead, or portions of them, standing singly or in rows in the most unexpected places least in the way in the crowded fields and gardens, awaiting removal to the final resting place. It is this custom, too, I am told, which has led to placing a large quantity of caustic lime in the bottom of the casket, on which the body rests, this acting as an effective absorbent.

Fig. 29.—Temporary burial, coffin thatched with straw; graves on the higher land at the right in background.

It is the custom in some parts of China, if not in all, to periodically restore the mounds, maintaining their hight and size, as is seen in the next two illustrations, and to decorate these once in the year with flying streamers of colored paper, the remnants of which may be seen in both Figs. 30 and 31, set there as tokens that the paper money has been burned upon them and its essence sent up in the smoke for the maintenance of the spirits of their departed friends. We have our memorial day; they have for centuries observed theirs with religious fidelity.

The usual expense of a burial among the working people is said to be $100, Mexican, an enormous burden when the day's wage or the yearly earning of the family is considered and when there is added to this the yearly expense of

Fig. 30.—Grave mounds recently restored and bearing the streamer standards in token of memorial services.

ancestor worship. How such voluntary burdens are assumed by people under such circumstances is hard to understand. Missionaries assert it is fear of evil consequences in this life and of punishment and neglect in the hereafter that leads to assuming them. Is it not far more likely that such is the price these people are willing to pay for a good name among the living and because of their deep and lasting friendship for the departed? Nor does it seem at all strange that a kindly, warm-hearted people with strong filial affection should have reached, early in their long history, a belief in one spirit of the departed which hovers about the home, one which hovers about the grave and another which wanders abroad, for surely there are associations with each of these conditions which must long and forcefully awaken memories of friends gone. If this view is possible may not such ancestral worship be an index of qualities of character strongly fixed and of the highest worth which, when improvements come that may relieve the heavy burdens now carried, will only shine more brightly and count more for right living as well as comfort?

Even in our own case it will hardly be maintained that our burial customs have reached their best and final solution, for in all civilized nations they are unnecessarily expensive and far too cumbersome. It is only necessary to mentally add the accumulation of a few centuries to our cemeteries to realize how impossible our practice must become. Clearly there is here a very important line for betterment which all nationalities should undertake.

When the steamer anchored at Shanghai the day was pleasant and the rain coats which greeted us in Yokohama were not in evidence but the numbers who had met the steamer in the hope of an opportunity for earning a trifle was far greater and in many ways in strong contrast with the Japanese. We were much surprised to find the men of so large stature, much above the Chinese usually seen in the United States. They were fully the equal of large Americans in frame but quite without surplus flesh yet

Fig. 31.—Group of grass-grown grave mounds carrying the streamer standards and showing the extensive occupation of land.

few appeared underfed. To realize that these are strong, hardy men it was only necessary to watch them carrying on their shoulders bales of cotton between them, supported by a strong bamboo; while the heavy loads they transport on wheel-barrows through the country over long distances,

Fig. 32.—Men freighters going inland with loads of matches.

as seen in Fig. 32, prove their great endurance. This same type of vehicle, too, is one of the common means of transporting people, especially Chinese women, and four, six and even eight may be seen riding together, propelled by a single wheelbarrow man.

III.

TO HONGKONG AND CANTON.

We had come to learn how the old-world farmers had been able to provide materials for food and clothing on such small areas for so many millions, at so low a price, during so many centuries, and were anxious to see them at the soil and among the crops. The sun was still south of the equator, coming north only about twelve miles per day, so, to save time, we booked on the next steamer for Hongkong to meet spring at Canton, beyond the Tropic of Cancer, six hundred miles farther south, and return with her.

On the morning of March 4th the Tosa Maru steamed out into the Yangtse river, already flowing with the increased speed of ebb tide. The pilots were on the bridge to guide her course along the narrow south channel through waters seemingly as brown and turbid as the Potomac after a rain. It was some distance beyond Gutzlaff Island, seventy miles to sea, where there is a lighthouse and a telegraph station receiving six cables, that we crossed the front of the out-going tide, showing in a sharp line of contrast stretching in either direction farther than the eye could see, across the course of the ship and yet it was the season of low water in this river. During long ages this stream of mighty volume has been loading upon itself in far-away Tibet, without dredge, barge, fuel or human effort, unused and there unusable soils, bringing them down from inaccessible hights across two or three thousand miles, building up with them, from under the sea, at the gateways of commerce, miles upon miles of the world's most fertile fields and gardens. Today on this river, winding through six hundred miles of the most highly cultivated fields, laid out on river-built plains, go large ocean steamers to the city of Hankow-Wuchang-Hanyang where 1,770,000 people live and trade within a radius less than four miles; while smaller steamers push on a thousand miles and are then but 130 feet above sea level.

Even now, with the aid of current, tide and man, these brown turbid waters are rapidly adding fertile delta plains for new homes. During the last twenty-five years Chung-ming island has grown in length some 1800 feet per year and today a million people are living and growing rice, wheat, cotton and sweet potatoes on 270 square miles of fertile plain where five hundred years ago were only submerged river sands and silt. Here 3700 people per square mile have acquired homes.

The southward voyage was over a quiet sea and as we passed among and near the off-shore islands these, as seen in Japan, appeared destitute of vegetation other than the low herbaceous types with few shrubs and almost no forest growth and little else that gave the appearance of green. Captain Harrison informed me that at no time in the year are these islands possessed of the grass-green verdure so often seen in northern climates, and yet the islands lie in a region of abundant summer rain, making it hard to understand why there is not a more luxuriant growth.

Sunday morning, March 7th, passing first extensive sugar refineries, found us entering the long, narrow and beautiful harbor of Hongkong. Here, lying at anchor in the ten square miles of water, were five battleships, several large ocean steamers, many coastwise vessels and a multitude of smaller craft whose yearly tonnage is twenty to thirty millions. But the harbor lies in the track of the terrible East Indian typhoon and, although sheltered on the north shore of a high island, one of these storms recently sunk nine vessels, sent twenty-three ashore, seriously damaged twenty-one others, wrought great destruction among the smaller craft and over a thousand dead were recovered.

Such was the destruction wrought by the September storm of 1906.

Our steamer did not go to dock but the Nippon Yusen Kaisha's launch transferred us to a city much resembling Seattle in possessing a scant footing between a long sea front and high steep mountain slopes behind. Here cliffs too steep to climb rise from the very sidewalk and are covered with a great profusion and variety of ferns, small bamboo, palms, vines, many flowering shrubs, all interspersed with pine and great banyan trees that do so much toward adding the beauty of northern landscapes to the tropical features which reach upward until hidden in a veil of fog that hung, all of the time we were there, over the city, over the harbor and stretched beyond Old and New Kowloon.

Hongkong island is some eleven miles long and but two to five miles wide, while the peak carrying the signal staff rises 1,825 feet above the streets from which ascends the Peak tramway, where, hanging from opposite ends of a strong cable, one car rises up the slope and another descends every fifteen to twenty minutes, affording communication with business houses below and homes in beautiful surroundings and a tempered climate above. Extending along the slopes of the mountains, too, above the city, are very excellent roads, carefully graded, provided with concrete gutters and bridges, along which one may travel on foot, on horseback, by ricksha or sedan chair, but too narrow for carriages. Over one of these we ascended along one side of Happy Valley, around its head and down the other side. Only occasionally could we catch glimpses of the summit through the lifting fog but the views, looking down and across the city and beyond the harbor with its shipping, and up and down the many ravines from viaducts, are among the choicest and rarest ever made accessible to the residents of any city. It was the beginning of the migratory season for birds, and trees and shrubbery thronged with many species.

Many of the women in Hongkong were seen engaged in such heavy manual labor with the men as carrying crushed rock and sand, for concrete and macadam work, up the steep street slopes long distances from the dock, but they were neither tortured nor incapacitated by bound feet. Like the men, they were of smaller stature than most seen at Shanghai and closely resemble the Chinese in the United States. Both sexes are agile, wiry and strong. Here we first saw lumber sawing in the open streets after the manner shown in Fig. 33, where wide boards were

Fig. 33.—Usual method of sawing lumber in China.

being cut from camphor logs. In the damp, already warm weather the men were stripped to the waist, their limbs bare to above the knee, and each carried a large towel for wiping away the profuse perspiration.

It was here, too, that we first met the remarkable staging
for the erection of buildings of four and six stories, set up
without saw, hammer or nail; without injury to or waste
of lumber and with the minimum of labor in construction
and removal. Poles and bamboo stems were lashed together
with overlapping ends, permitting any interval or hight
to be secured without cutting or nailing, and admitting of
ready removal with absolutely no waste, all parts being
capable of repeated use unless it be some of the materials
employed in tying members. Up inclined stairways, from
staging to staging, in the erection of six-story granite buildings, mortar was being carried in baskets swinging from

Fig. 34.—Happy Valley, Hongkong Island, with its terraced gardens and
scattered dwellings.

bamboo poles on the shoulders of men and women, as the
cheapest hoists available in English Hongkong where there
is willing human labor and to spare.

The Singer sewing machine, manufactured in New Jersey,

was seen in many Chinese shops in Hongkong and other cities, operated by Chinese men and women, purchased, freight prepaid, at two-thirds the retail price in the United States. Such are the indications of profit to manufacturers on the home sale of home-made goods while at the same time reaping good returns from a large trade in heathen lands, after paying the freight.

Fig. 35.—Statuary floral pieces in florist's garden, Happy Valley, Hong-kong, China.

Industrial China, Korea and Japan do not observe our weekly day of rest and during our walk around Happy Valley on Sunday afternoon, looking down upon its terraced gardens and tiny fields, we saw men and women busy fitting the soil for new crops, gathering vegetables for market, feeding plants with liquid manure and even irrigating certain crops, notwithstanding the damp, foggy, showery weather. Turning the head of the valley, attention was drawn to a walled enclosure and a detour down

Fig 36.—A fair type of garden culture seen in Happy Valley, Hongkong.

the slope brought us to a florist's garden within which were rows of large potted foliage plants of semi-shrubbery habit, seen in Fig. 35, trained in the form of life-size human figures with limbs, arms and trunk provided with highly glazed and colored porcelain feet, hands and head. These, with many other potted plants and trees, including dwarf varieties, are grown under out-door lattice shelters in different parts of China, for sale to the wealthy Chinese families.

Fig. 37.—Receptacles for collecting liquid manure, and at their right a pile of ashes and a pile of stable manure for fertilizing the garden.

How thorough is the tillage, how efficient and painstaking the garden fitting, and how closely the ground is crowded to its upper limit of producing power are indicated in Fig. 36; and when one stops and studies the detail in such gardens he expects in its executor an orderly, careful, frugal and industrious man, getting not a little satisfaction out of his creations however arduous his task or prolonged his day. If he is in the garden or one meets him at the house, clad as the nature of his duties and compensation have determined, you may be disappointed or feel arising an unkind judgment. But who would risk a reputation so clad and so environed? Many were the times, during our walks in the fields and gardens among these old, much misunderstood, misrepresented and undervalued people, when the bond of common interest was recognized between us, that there showed through the face the spirit which put aside both dress and surroundings and the man stood forth who, with fortitude and rare wisdom, is feeding the millions and who has carried through centuries the terrible burden of taxes levied by dishonor and needless wars. Nay, more than this, the man stood forth who has kept alive the seeds of manhood and has nourished them into such sturdy stock as has held the stream of progress along the best interests of civilization in spite of the drift-wood heaped upon it.

Not only are these people extremely careful and painstaking in fitting their fields and gardens to receive the crop, but they are even more scrupulous in their care to make everything that can possibly serve as fertilizer for the soil, or food for the crop being grown, do so unless there is some more remunerative service it may render. Expense is incurred to provide such receptacles as are seen in Fig. 37 for receiving not only the night soil of the home and that which may be bought or otherwise procured, but in which may be stored any other fluid which can serve as plant food. On the right of these earthenware jars too is a pile of ashes and one of manure. All such materials are saved and used in the most advantageous ways to enrich the soil or to nourish the plants being grown.

Generally the liquid manures must be diluted with water to a greater or less extent before they are "fed", as the Chinese say, to their plants, hence there is need of an abundant and convenient water supply. One of these is seen in Fig. 38, where the Chinaman has adopted the modern galvanized iron pipe to bring water from the mountain

Fig. 38.—Water brought from mountain side in three-fourths inch iron pipe to use in diluting liquid manure and in garden irrigation.

slope of Happy Valley to his garden. By the side of this tank are the covered pails in which the night soil was brought, perhaps more than a mile, to be first diluted and then applied. But the more general method for supplying water is that of leading it along the ground in channels or ditches to a small reservoir in one corner of a terraced field or garden, as seen in Fig. 39, where it is held and the surplus led down from terrace to terrace, giving each its permanent supply. At the upper right corner of the engraving may be seen two manure receptacles and a third stands near the reservoir. The plants on the lower terrace are water cress and those above the same. At this time of the year, on the terraced gardens of Happy Valley, this is one of the crops most extensively grown.

Walking among these gardens and isolated homes, we passed a pig pen provided with a smooth, well-laid stone floor that had just been washed scrupulously clean, like the floor of a house. While I was not able to learn other facts regarding this case, I have little doubt that the washings from this floor had been carefully collected and taken to some receptacle to serve as a plant food.

Looking backward as we left Hongkong for Canton on the cloudy evening of March 8th, the view was wonderfully beautiful. We were drawing away from three cities, one, electric-lighted Hongkong rising up the steep slopes, suggesting a section of sky set with a vast array of stars of all magnitudes up to triple Jupiters; another, old and new Kowloon on the opposite side of the harbor; and between these two, separated from either shore by wide reaches of wholly unoccupied water, lay the third, a mid-strait city of sampans, junks and coastwise craft of many kinds segregated, in obedience to police regulation, into blocks and streets with each setting sun, but only to scatter again with the coming morn. At night, after a fixed hour, no one is permitted to leave shore and cross the vacant water strip except from certain piers and with the permission of the police, who take the number of the sampan and the names of its occupants. Over the harbor three large search

Fig. 39.—Series of terraces showing small water reservoir near center, three receptacles for liquid manure, and a bed of water cress in the foreground.

lights were sweeping and it was curious to see the junks and other craft suddenly burst into full blazes of light, like so many monstrous fire-flies, to disappear and reappear as the lights came and went. Thus is the mid-strait city lighted and policed and thus have steps been taken to lessen the number of cases of foul play where people have left the wharves at night for some vessel in the strait, never to be heard from again.

Some ninety miles is the distance by water to Canton, and early the next morning our steamer dropped anchor off the foreign settlement of Shameen. Through the kindness of Consul-General Amos P. Wilder in sending a telegram to the Canton Christian College, their little steam launch met the boat and took us directly to the home of the college on Honam Island, lying in the great delta south of the city where sediments brought by the Si-kiang—west, Pei-kiang—north, and Tung-kiang—east—rivers through long centuries have been building the richest of land, which, because of the density of population, are squared up everywhere to the water's edge and appropriated as fast as formed, and made to bring forth materials for food, fuel and raiment in vast quantities.

It was on Honam Island that we walked first among the grave lands and came to know them as such, for Canton Christian College stands in the midst of graves which, although very old, are not permitted to be disturbed and the development of the campus must wait to secure permission to remove graves, or erect its buildings in places not the most desirable. Cattle were grazing among the graves and with them a flock of some 250 of the brown Chinese geese, two-thirds grown, was watched by boys, gleaning their entire living from the grave lands and adjacent water. A mature goose sells in Canton for $1.20, Mexican, or less than 52 cents, gold, but even then how can the laborer whose day's wage is but ten or fifteen cents afford one for his family? Here, too, we saw the Chinese persistent, never-ending industry in keeping their land, their sunshine and their rain, with themselves, busy in producing something needful. Fields which had matured two crops of rice during the long summer, had been laboriously, and largely by hand labor, thrown into strong ridges as seen in Fig. 40, to permit still a third winter crop of some vegetable to be taken from the land.

Fig. 40.—Looking across fields which have borne two crops of rice, now ridged for leeks and other vegetables as a winter crop.

But this intensive, continuous cropping of the land spells soil exhaustion and creates demands for maintenance and restoration of available plant food or the adding of large quantities of something quickly convertible into it, and so here in the fields on Honam Island, as we had found in Happy Valley, there was abundant evidence of the most careful attention and laborious effort devoted to plant feeding. The boat standing in the canal in Fig. 41 had come from Canton in the early morning with two tons of human manure and men were busy applying it, in diluted form, to beds of leeks at the rate of 16,000 gallons per acre, all carried on the shoulders in such pails as stand in the foreground. The material is applied with long-handled dippers holding a gallon, dipping it from the pails, the men wading, with bare feet and trousers rolled above the knees, in the water of the furrows between the beds. This is one of their ways of "feeding the crop," and they have other methods of "manuring the soil."

Fig. 41.—Boat load of human waste in canal on Honam Island, brought from Canton and being used in feeding winter vegetables.

One of these we first met on Honam Island. Large amounts of canal mud are here collected in boats and brought to the fields to be treated and there left to drain and dry before distributing. Both the material used to feed the crop and that used for manuring the land are waste products, hindrances to the industry of the region, but the Chinese make them do essential duty in maintaining its life. The human waste must be disposed of. They return it to the soil. We turn it into the sea. Doing so, they save for plant feeding more than a ton of phosphorus (2712 pounds) and more than two tons of potassium (4488 pounds) per day for each million of adult population. The mud collects in their canals and obstructs movement. They must be kept open. The mud is highly charged with organic matter and would add humus to the soil if applied to the fields, at the same time raising their level above the river and canal, giving them better drainage; thus are they turning to use what is otherwise waste, causing the labor which must be expended in disposal to count in a remunerative way.

During the early morning ride to Canton Christian College and three others which we were permitted to enjoy in the launch on the canal and river waters, everything was again strange, fascinating and full of human interest. The Cantonese water population was a surprise, not so much for its numbers as for the lithe, sinewy forms, bright eyes and cheerful faces, particularly among the women, young and old. Nearly always one or more women, mother and daughter oftenest, grandmother many times, wrinkled, sometimes grey, but strong, quick and vigorous in motion, were manning the oars of junks, houseboats and sampans. Sometimes husband and wife and many times the whole family were seen together when the craft was both home and business boat as well. Little children were gazing from most unexpected peek holes, or they toddled tethered from a waist belt at the end of as much rope as would arrest them above water, should they go overboard. And the cat was similarly tied. Through an overhanging latticed stern, too, hens craned their necks, longing for scenes they could not reach. With bare heads, bare feet, in short trousers and all dressed much alike, men, women, boys and girls showed equal mastery of the oar. Beginning so young, day and night in the open air on the tide-swept streams and canals, exposed to all of the sunshine the fogs and clouds will permit, and removed from the dust and filth of streets, it would seem that if the children survive at all they must develop strong. The appearance of the women somehow conveyed the impression that they were more vigorous and in better fettle than the men.

Boats selling many kinds of steaming hot dishes were common. Among these was rice tied in green leaf wrappers, three small packets in a cluster suspended by a strand of some vegetable fiber, to be handed hot from the cooker to the purchaser, some one on a passing junk or on an in-coming or out-going boat. Another would buy hot water for a brew of tea, while still another, and for a single cash, might be handed a small square of cotton cloth, wrung hot from the water, with which to wipe his face and hands and then be returned.

Perhaps nothing better measures the intensity of the maintenance struggle here, and better indicates the minute economies practiced, than the value of their smallest currency unit, the Cash, used in their daily retail transactions. On our Pacific coast, where less thought is given to little economies than perhaps anywhere else in the world, the nickel is the smallest coin in general use, twenty to the dollar. For the rest of the United States and in most English speaking countries one hundred cents or half pennies measure an equal value. In Russia 170 kopecks, in Mexico 200 centavos, in France 250 two-centime pieces, and in Austria-Hungary 250 two-heller coins equal the United States dollar; while in Germany 400 pfennigs, and in India 400 pie are required for an equal value. Again 500 penni in Finland and of stotinki in Bulgaria, of centesimi in Italy and of half cents in Holland equal our dollar; but in China the small daily financial transactions are measured against a much smaller unit, their Cash, 1500 to 2000 of which are required to equal the United States dollar, their purchasing power fluctuating daily with the price of silver.

In the Shantung province, when we inquired of the farmers the selling prices of their crops, their replies were given like this: "Thirty-five strings of cash for 420 catty of wheat and twelve to fourteen strings of cash for 1000 catty of wheat straw." At this time, according to my interpreter, the value of one string of cash was 40 cents Mexican, from which it appears that something like 250 of these coins were threaded on a string. Twice we saw a wheelbarrow heavily loaded with strings of cash being transported through the streets of Shanghai, lying exposed on the frame, suggesting chains of copper more than money. At one of the go-downs or warehouses in Tsingtao, where freight was being transferred from a steamer, the carriers were receiving their pay in these coin. The pay-master stood in the doorway with half a bushel of loose cash in a grain sack at his feet. With one hand he received the bamboo tally-sticks from the stevedores and with the other paid the cash for service rendered.

Reference has been made to buying hot water. In a sampan managed by a woman and her daughter, who took us ashore, the middle section of the boat was furnished in the manner of a tiny sitting-room, and on the sideboard sat the complete embodiment of our fireless cookers, keeping boiled water hot for making tea. This device and the custom are here centuries old and throughout these countries boiled water, as tea, is the universal drink, adopted no doubt as a preventive measure against typhoid fever and allied diseases. Few vegetables are eaten raw and nearly all foods are taken hot or recently cooked if not in some way pickled or salted. Houseboat meat shops move among the many junks on the canals. These were provided with a compartment communicating freely with the canal water where the fish were kept alive until sold. At the street markets too, fish are kept alive in large tubs of water systematically aerated by the water falling from an elevated receptacle in a thin stream. A live fish may even be sliced before the eyes of a purchaser and the unsold portion returned to the water. Poultry is largely retailed alive although we saw much of it dressed and cooked to a uniform rich brown, apparently roasted, hanging exposed in the markets of the very narrow streets in Canton, shaded from the hot sun under awnings admitting light overhead through translucent oyster-shell latticework. Perhaps these fowl had been cooked in hot oil and before serving would be similarly heated. At any rate it is per-

fectly clear that among these people many very fundamental sanitary practices are rigidly observed.

One fact which we do not fully understand is that,
wherever we went, house flies were very few. We never
spent a summer with so little annoyance from them as this
one in China, Korea and Japan. It may be that our
experience was exceptional but, if so, it could not be
ascribed to the season of our visit for we have found flies
so numerous in southern Florida early in April as to
make the use of the fly brush at the table very necessary.
If the scrupulous husbanding of waste refuse so universally
practiced in these countries reduces the fly nuisance and
this menace to health to the extent which our experience
suggests, here is one great gain. We breed flies in countless
millions each year, until they become an intolerable nuisance, and then expend millions of dollars on screens and
fly poison which only ineffectually lessen the intensity and
danger of the evil.

The mechanical appliances in use on the canals and
in the shops of Canton demonstrate that the Chinese
possess constructive ability of a high order, notwithstanding
so many of these are of the simplest forms. This statement is well illustrated in the simple yet efficient foot-
power seen in Fig. 42, where a father and his two sons
are driving an irrigation pump, lifting water at the rate
of seven and a half acre-inches per ten hours, and at a
cost, including wage and food, of 36 to 45 cents, gold.
Here, too, were large stern-wheel passenger boats, capable
of carrying thirty to one hundred people, propelled by
the same foot-power but laid crosswise of the stern, the
men working in long single or double lines, depending on
the size of the boat. On these the fare was one cent,
gold, for a fifteen mile journey, a rate one-thirtieth our
two-cent railway tariff. The dredging and clearing of
the canals and water channels in and about Canton
is likewise accomplished with the same foot-power,
often by families living on the dredge boats. A
dipper dredge is used, constructed of strong bamboo

strips woven into the form of a sliding, two-horse road scraper, guided by a long bamboo handle. The dredge is drawn along the bottom by a rope winding about the projecting axle of the foot-power, propelled by three or more people. When the dipper reaches the axle and is raised from the water it is swung aboard, emptied and returned by means of a long arm like the old well sweep,

Fig. 42.—The wooden foot-power of China, being used to propel the wooden-chain irrigation pump.

operated by a cord depending from the lower end of the lever, the dipper swinging from the other. Much of the mud so collected from the canals and channels of the city is taken to the rice and mulberry fields, many square miles of which occupy the surrounding country. Thus the channels are kept open, the fields grow steadily higher above flood level, while their productive power is maintained by the plant food and organic matter carried in the sediment.

The mechanical principle involved in the boy's button buzz was applied in Canton and in many other places for operating small drills as well as in grinding and polishing appliances used in the manufacture of ornamental ware. The drill, as used for boring metal, is set in a straight shaft, often of bamboo, on the upper end of which is mounted a circular weight. The drill is driven by a pair of strings with one end attached just beneath the momentum weight and the other fastened at the ends of a cross hand-bar, having a hole at its center through which the shaft carrying the drill passes. Holding the drill in position for work and turning the shaft, the two cords are wrapped about it in such a manner that simple downward pressure on the hand bar held in the two hands unwinds the cords and thus revolves the drill. Relieving the pressure at the proper time permits the momentum of the revolving weight to rewind the cords and the next downward pressure brings the drill again into service.

IV

UP THE SI-KIANG, WEST RIVER.

On the morning of March 10th we took passage on the Nanning for Wuchow, in Kwangsi province, a journey of 220 miles up the West river, or Sikiang. The Nanning is one of two English steamers making regular trips between the two places, and it was the sister boat which in the summer of 1906 was attacked by pirates on one of her trips and all of the officers and first class passengers killed while at dinner. The cause of this attack, it is said, or the excuse for it, was threatened famine resulting from destructive floods which had ruined the rice and mulberry crops of the great delta region and had prevented the carrying of manure and bean cake as fertilizers to the tea fields in the hill lands beyond, thus bringing ruin to three of the great staple crops of the region. To avoid the recurrence of such tragedies the first class quarters on the Nanning had been separated from the rest of the ship by heavy iron gratings thrown across the decks and over the hatchways. Armed guards stood at the locked gateways, and swords were hanging from posts under the awnings of the first cabin quarters, much as saw and ax in our passenger coaches. Both British and Chinese gunboats were patrolling the river; all Chinese passengers were searched for concealed weapons as they came aboard, even though Government soldiers, and all arms taken into custody until the end of the journey. Several of the large Chinese merchant junks which were passed, carrying valuable cargoes on the river, were armed with small cannon, and when riding by rail from Canton to Sam Shui, a government pirate detective was in our coach.

The Sikiang is one of the great rivers of China and indeed of the world. Its width at Wuchow at low water was nearly a mile and our steamer anchored in twenty-four feet of water to a floating dock made fast by huge iron chains reaching three hundred feet up the slope to the city proper, thus providing for a rise of twenty-six feet in the river at its flood stage during the rainy season. In a narrow section of river where it winds through Shui Hing gorge, the water at low stage has a depth of more than twenty-five fathoms, too deep for anchorage, so in times of prospective fog, boats wait for clearing weather. Fluctuations in the hight of the river limit vessels passing up to Wuchow to those drawing six and a half feet of water during the low stage, and at high stage to those drawing sixteen feet.

When the West river emerges from the high lands, with its burden of silt, to join its waters with those of the North and East rivers, it has entered a vast delta plain some eighty miles from east to west and nearly as many from north to south, and this has been canalized, diked, drained and converted into the most productive of fields, bearing three or more crops each year. As we passed westward through this delta region the broad flat fields, surrounded by dikes to protect them against high water, were being plowed and fitted for the coming crop of rice. In many places the dikes which checked off the fields were planted with bananas and in the distance gave the appearance of extensive orchards completely occupying the ground. Except for the water and the dikes it was easy to imagine that we were traversing one of our western prairie sections in the early spring, at seeding time, the scattered farm villages here easily suggested distant farmsteads; but a nearer approach to the houses showed that the roofs and sides were thatched with rice straw and stacks were very numerous about the buildings. Many tide gates were set in the dikes, often with double trunks.

At times we approached near enough to the fields to see how they were laid out. From the gates long canals, six to eight feet wide, led back sometimes eighty or a hundred rods. Across these and at right angles, head channels were cut and between them the fields were plowed in long straight lands some two rods wide, separated by water furrows. Many of the fields were bearing sugar cane standing eight feet high. The Chinese do no sugar refining but boil the sap until it will solidify, when it is run into cakes resembling chocolate or our brown maple sugar. Immense quantities of sugar cane, too, are exported to the northern provinces, in bundles wrapped with matting or other cover, for the retail markets where it is sold, the canes being cut in short sections and sometimes peeled, to be eaten from the hands as a confection.

Much of the way this water-course was too broad to permit detailed study of field conditions and crops, even with a glass. In such sections the recent dikes often have the appearance of being built from limestone blocks but a closer view showed them constructed from blocks of the river silt cut and laid in walls with slightly sloping faces. In time however the blocks weather and the dikes become rounded earthen walls.

We passed two men in a boat, in charge of a huge flock of some hundreds of yellow ducklings. Anchored to the bank was a large houseboat provided with an all-around, over-hanging rim and on board was a stack of rice straw and other things which constituted the floating home of the ducks. Both ducks and geese are reared in this manner in large numbers by the river population. When it is desired to move to another feeding ground a gang plank is put ashore and the flock come on board to remain for the night or to be landed at another place.

About five hours journey westward in this delta plain, where the fields lie six to ten feet above the present water stage, we reached the mulberry district. Here the plants are cultivated in rows about four feet apart, having the habit of small shrubs rather than of trees, and so much resembling cotton that our first impression was that we were in an extensive cotton district. On the lower lying areas, surrounded by dikes, some fields were laid out in the manner of the old Italian or English water meadows, with a shallow irrigation furrow along the crest of the bed and much deeper drainage ditches along the division line between them. Mulberries were occupying the ground before the freshly cut trenches we saw were dug, and all the

Fig. 43.—Field of mulberry having the surface covered with fresh earth taken from ditches dividing the land into beds.

surface between the rows had been evenly overlaid with the fresh earth removed with the spade, the soil lying in blocks essentially unbroken. In Fig. 43 may be seen the mulberry crop on a similarly treated surface, between Canton and Samshui, with the earth removed from the trenches laid evenly over the entire surface between and around the plants, as it came from the spade.

At frequent intervals along the river, paths and steps were seen leading to the water and within a distance of a quarter of a mile we counted thirty-one men and women carrying mud in baskets on bamboo poles swung across their shoulders, the mud being taken from just above the water line. The disposition of this material we could not see as it was carried beyond a rise in ground. We have little doubt that the mulberry fields were being covered with it. It was here that a rain set in and almost like magic the fields blossomed out with great numbers of giant rain hats and kittysols, where people had been unobserved before. From one o'clock until six in the afternoon we had travelled continuously through these mulberry fields stretching back miles from our line of travel on either hand, and the total acreage must have been very large. But we had now nearly reached the margin of the delta and the mulberries changed to fields of grain, beans, peas and vegetables.

After leaving the delta region the balance of the journey to Wuchow was through a hill country, the slopes rising steeply from near the river bank, leaving relatively little tilled or readily tillable land. Rising usually five hundred to a thousand feet, the sides and summits of the rounded, soil-covered hills were generally clothed with a short herbaceous growth and small scattering trees, oftenest pine, four to sixteen feet high, Fig. 44 being a typical landscape of the region.

In several sections along the course of this river there are limited areas of intense erosion where naked gulleys of no mean magnitude have developed but these were exceptions and we were continually surprised at the remarkable steepness of the slopes, with convexly rounded contours almost everywhere, well mantled with soil, devoid of gulleys and completely covered with herbaceous growth dotted with small trees. The absence of forest growth finds its explanation in human influence rather than natural conditions.

Throughout the hill-land section of this mighty river the most characteristic and persistent human features were the stacks of brush-wood and the piles of stove wood along the banks or loaded upon boats and barges for the market. The brush- wood was largely made from the boughs of pine, tied into bundles and stacked like grain. The stove wood was usually round, peeled and made from the limbs and trunks of trees two to five inches in diameter. All this fuel was coming to the river from the back country, sent down

Fig. 44.—Scantily wooded hills on the Sikiang. Boatload of stove wood; stack of pine bough fuel in bundles behind. The trees are small pine from which the lower limbs have been cut for fuel.

along steep slides which in the distance resemble paths leading over hills but too steep for travel. The fuel was loaded upon large barges, the boughs in the form of stacks to shed rain but with a tunnel leading into the house of the boat about which they were stacked, while the wood was similarly corded about the dwelling, as seen in Fig. 44. The wood was going to Canton and other delta cities while the pine boughs were taken to the lime and cement kilns, many of which were located along the river. Absolutely the whole tree, including the roots and the needles, is saved and burned; no waste is permitted.

The up-river cargo of the Nanning was chiefly matting rush, taken on at Canton, tied in bundles like sheaves of wheat. It is grown upon the lower, newer delta lands by methods of culture similar to those applied to rice, Fig. 45 showing a field as seen in Japan.

The rushes were being taken to one of the country villages on a tributary of the Sikiang and the steamer was met by a flotilla of junks from this village, some forty-five miles up the stream, where the families live who do the weaving. On the return trip the flotilla again met the steamer with a cargo of the woven matting. In keeping record of packages transferred the Chinese use a simple and unique method. Each carrier, with his two bundles, received a pair of tally sticks. At the gang-plank sat a man with a tally-case divided into twenty compartments, each of which could receive five, but no more, tallies. As the bundles left the steamer the tallies were placed in the tally-case until it contained one hundred, when it was exchanged for another.

Wuchow is a city of some 65,000 inhabitants, standing back on the higher ground, not readily visible from the steamer landing nor from the approach on the river. On the foreground, across which stretched the anchor chains of the dock, was living a floating population, many in shelters less substantial than Indian wigwams, but engaged in a great variety of work, and many water buffalo had been tied for the night along the anchor chains. Before July much of this area would lie beneath the flood waters of the Sikiang.

Here a ship builder was using his simple, effective bow-brace, boring holes for the dowel pins in the planking for his ship, and another was bending the plank to the proper curvature. The bow-brace consisted of a bamboo stalk carrying the bit at one end and a shoulder rest at the other. Pressing the bit to its work with the shoulder, it was driven

Fig. 45.—Landscape in Japan showing fields of rice and matting rush. The dark area in the foreground and others back are occupied by the rush.

with the string of a long bow wrapped once around the
stalk by drawing the bow back and forth, thus rapidly and
readily revolving the bit.

The bending of the long, heavy plank, four inches thick
and eight inches wide, was more simple still. It was saturated with water and one end raised on a support four feet
above the ground. A bundle of burning rice straw moved

Fig. 46.—Wooden fork shaped from the limbs of a tree by simple means of
steaming and drying.

along the under side against the wet wood had the effect
of steaming the wood and the weight of the plank caused
it to gradually bend into the shape desired. Bamboo poles
are commonly bent or straightened in this manner to suit
any need and Fig. 46 shows a wooden fork shaped in the
manner described from a small tree having three main
branches. This fork is in the hands of my interpreter and
was used by the woman standing at the right, in turning
wheat.

When the old ship builder had finished shaping his plank
he sat down on the ground for a smoke. His pipe was one

joint of bamboo stem a foot long, nearly two inches in diameter and open at one end. In the closed end, at one side, a small hole was bored for draft. A charge of tobacco was placed in the bottom, the lips pressed into the open end and the pipe lighted by suction, holding a lighted match at the small opening. To enjoy his pipe the bowl rested on the ground between his legs. With his lips in the bowl and a long breath, he would completely fill his lungs, retaining the smoke for a time, then slowly expire and fill the lungs again, after an interval of natural breathing.

On returning to Canton we went by rail, with an interpreter, to Samshui, visiting fields along the way, and Fig. 47 is a view of one landscape. The woman was picking roses among tidy beds of garden vegetables. Beyond her and in front of the near building are two rows of waste receptacles. In the center background is a large "go-down," in function that of our cold storage warehouse and in part that of our grain elevator for rice. In them, too, the wealthy store their fur-lined winter garments for safe keeping. These are numerous in this portion of China and the rank of a city is indicated by their number. The conical hillock is a large near-by grave mound and many others serrate the sky line on the hill beyond.

In the next landscape, Fig. 48, a crop of winter peas, trained to canes, are growing on ridges among the stubble of the second crop of rice. In front is one canal, the double ridge behind is another and a third canal extends in front of the houses. Already preparations were being made for the first crop of rice, fields were being flooded and fertilized. One such is seen in Fig. 49, where a laborer was engaged at the time in bringing stable manure, wading into the water to empty the baskets.

Two crops of rice are commonly grown each year in southern China and during the winter and early spring, grain, cabbage, rape, peas, beans, leeks and ginger may occupy the fields as a third or even fourth crop, making the total year's product from the land very large; but the amount of thought, labor and fertilizers given to securing

Fig. 47.—Landscape at Samshui, near Canton, China.

Fig. 48.—Peas grown in winter after second crop of rice; with three parallel canals.

these is even greater and beyond anything Americans will endure. How great these efforts are will be appreciated from what is seen in Fig. 50, representing two fields thrown into high ridges, planted to ginger and covered with straw. All of this work is done by hand and when the time for rice planting comes every ridge will again be thrown down and the surface smoothed to a water level. Even when the ridges and beds are not thrown down for the crops of rice, the furrows and the beds will change places so that all the soil is worked over deeply and mainly through hand labor. The statement so often made, that these people only barely scratch the surface of their fields with the crudest of tools is very far from the truth, for their soils are worked deeply and often, notwithstanding the fact that their plowing, as such, may be shallow.

Through Dr. John Blumann of the missionary hospital at Tungkun, east from Canton, we learned that the good rice lands there a few years ago sold at $75 to $130 per acre but that prices are rising rapidly. The holdings of the better class of farmers there are ten to fifteen mow,—one and two-thirds to two and a half acres—upon which are maintained families numbering six to twelve. The day's wage of a carpenter or mason is eleven to thirteen cents of our currency, and board is not included, but a day's ration for a laboring man is counted worth fifteen cents, Mexican, or less than seven cents, gold.

Fish culture is practiced in both deep and shallow basins, the deep permanent ones renting as high as $30 gold, per acre. The shallow basins which can be drained in the dry season are used for fish only during the rainy period, being later drained and planted to some crop. The permanent basins have often come to be ten or twelve feet deep, increasing with long usage, for they are periodically drained by pumping and the foot or two of mud which has accumulated, removed and sold as fertilizer to planters of ri e and other crops. It is a common practice, too, among the fish growers, to fertilize the ponds, and in case a foot path leads alongside, screens are built over the water to provide

Fig. 49.—Flooded fields being fertilized preparatory for rice; winter peas and beans beyond, with dwellings in the background.

accommodation for travelers. Fish reared in the better
fertilized ponds bring a higher price in the market. The
fertilizing of the water favors a stronger growth of food
forms, both plant and animal, upon which the fish live and

Fig. 50.—Fields of ginger just planted; ridged and furrowed for drainage,
showing the amount of hand labor performed to secure the winter crop,
following two of rice.

they are better nourished, making a more rapid growth,
giving their flesh better qualities, as is the case with well
fed animals.

In the markets where fish are exposed for sale they are
often sliced in halves lengthwise and the cut surface

smeared with fresh blood. In talking with Dr. Blumann as to the reason for this practice he stated that the Chinese very much object to eating meat that is old or tainted and that he thought the treatment simply had the effect of making the fish look fresher. I question whether this treatment with fresh blood may not have a real antiseptic effect and very much doubt that people so shrewd as the Chinese would be misled by such a ruse.

V.

EXTENT OF CANALIZATION AND SURFACE FITTING OF FIELDS.

On the evening of March 15th we left Canton for Hong-kong and the following day embarked again on the Tosa Maru for Shanghai. Although our steamer stood so far to sea that we were generally out of sight of land except for some off-shore islands, the water was turbid most of the way after we had crossed the Tropic of Cancer off the mouth of the Han river at Swatow. Over a sea bottom measuring more than six hundred miles northward along the coast, and perhaps fifty miles to sea, unnumbered acre-feet of the richest soil of China are being borne beyond the reach of her four hundred millions of people and the children to follow them. Surely it must be one of the great tasks of future statesmanship, education and engineering skill to divert larger amounts of such sediments close along inshore in such manner as to add valuable new land annually to the public domain, not alone in China but in all countries where large resources of this type are going to waste.

In the vast Cantonese delta plains which we had just left, in the still more extensive ones of the Yangtse kiang to which we were now going, and in those of the shifting Hwang ho further north, centuries of toiling millions have executed works of almost incalculable magnitude, fundamentally along such lines as those just suggested. They have accomplished an enormous share of these tasks by sheer force of body and will, building levees, digging ca-

7

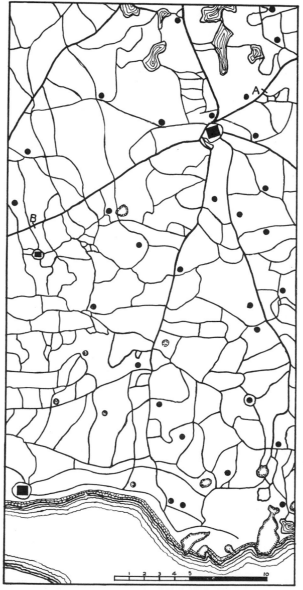

Fig. 51.—Map of main canals in 718 square miles of Chekiang Province. Each line represents a canal.

nals, diverting the turbid waters of streams through them
and then carrying the deposits of silt and organic growth
out upon the fields, often borne upon the shoulders of men
in the manner we have seen.

It is well nigh impossible, by word or map, to convey an
adequate idea of the magnitude of the systems of canalization and delta and other lowland reclamation work, or of
the extent of surface fitting of fields which have been effected in China, Korea and Japan through the many centuries, and which are still in progress. The lands so reclaimed and fitted constitute their most enduring asset and
they support their densest populations. In one of our
journeys by houseboat on the delta canals between Shanghai and Hangchow, in China, over a distance of 117 miles,
we made a careful record of the number and dimensions of
lateral canals entering and leaving the main one along
which our boat-train was traveling. This record shows
that in 62 miles, beginning north of Kashing and extending south to Hangchow, there entered from the west 134
and there left on the coast side 190 canals. The average
width of these canals, measured along the water line, we
estimated at 22 and 19 feet respectively on the two sides.
The hight of the fields above the water level ranged from
four to twelve feet, during the April and May stage of
water. The depth of water, after we entered the Grand
Canal, often exceeded six feet and our best judgment would
place the average depth of all canals in this part of China
at more than eight feet below the level of the fields.

In Fig. 51, representing an area of 718 square miles in
the region traversed, all lines shown are canals, but scarcely more than one-third of those present are shown on the
map. Between A, where we began our records, before
reaching Kashing, and B, near the left margin of the map,
there were forty-three canals leading in from the up-
country side, instead of the eight shown, and on the coast
side there were eighty-six leading water out into the delta
plain toward the coast, instead of the twelve shown. Again,
on one of our trips by rail, from Shanghai to Nanking, we

made a similar record of the number of canals seen from the train, close along the track, and the notes show, in a distance of 162 miles, 593 canals between Lungtan and Nansiang. This is an average of more than three canals per mile for this region and that between Shanghai and Hangchow.

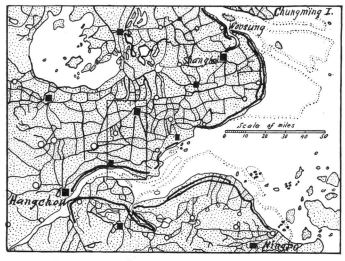

Fig. 52.—Sketch map of portions of Chekiang and Kiangsu Provinces, representing some 2,700 miles of main canals and over 300 miles of sea-wall. The sea-walls are represented by the very heavy black lines. The small rectangle shows the area covered by Fig. 51.

The extent, nature and purpose of these vast systems of internal improvement may be better realized through a study of the next two sketch maps. The first, Fig. 52, represents an area 175 by 160 miles, of which the last illustration is the portion enclosed in the small rectangle. On this area there are shown 2,700 miles of canals and only about one-third of the canals shown in Fig. 51 are laid down on this map, and according to our personal observations there are three times as many canals as are shown on the map of which Fig. 51 represents a part. It is probable, therefore, that there exists today in the area of Fig. 52 not less than 25,000 miles of canals.

In the next illustration, Fig. 53, an area of northeast China, 600 by 725 miles, is represented. The unshaded land area covers nearly 200,000 square miles of alluvial plain. This plain is so level that at Ichang, nearly a thousand miles up the Yangtse, the elevation is only 130 feet above the sea. The tide is felt on the river to beyond Wuhu, 375 miles from the coast. During the summer the depth of water in the Yangtse is sufficient to permit ocean vessels drawing twenty-five feet of water to ascend six hundred miles to Hankow, and for smaller steamers to go on to Ichang, four hundred miles further.

The location, in this vast low delta and coastal plain, of the system of canals already described, is indicated by the two rectangles in the south-east corner of the sketch map, Fig. 53. The heavy barred black line extending from Hangchow in the south to Tientsin in the north represents the Grand Canal which has a length of more than eight hundred miles. The plain, east of this canal, as far north as the mouth of the Hwang ho in 1852, is canalized much as is the area shown in Fig. 52. So, too, is a large area both sides of the present mouth of the same river in Shantung and Chihli, between the canal and the coast. Westward, up theYangtse valley, the provinces of Anhwei, Kiangsi, Hunan and Hupeh have very extensive canalized tracts, probably exceeding 28,000 square miles in area, and Figs. 54 and 55 are two views in this more western region. Still further west, in Szechwan province, is the Chengtu plain, thirty by seventy miles, with what has been called "the most remarkable irrigation system in China."

Westward beyond the limits of the sketch map, up the Hwang ho valley, there is a reach of 125 miles of irrigated lands about Ninghaifu, and others still farther west, at Lanchowfu and at Suchow where the river has attained an elevation of 5,000 feet, in Kansu province; and there is still to be named the great Canton delta region. A conservative estimate would place the miles of canals and leveed rivers in China, Korea and Japan equal to eight times the number represented in Fig. 52. Fully 200,000 miles in all. Forty canals across the United States from east to west and sixty from north to south would not equal, in number

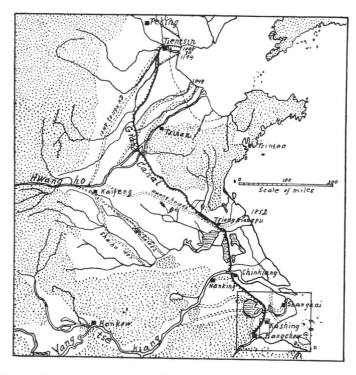

Fig. 53.—Sketch map of northeast China showing the alluvial plain and the Grand Canal, extending 800 miles through it from Hangchow to Tientsin. The unshaded land area lies mostly less than 100 feet above sea level.

of miles those in these three countries today. Indeed, it is probable that this estimate is not too large for China alone.

As adjuncts to these vast canalization works there have been enormous amounts of embankment, dike and

Fig. 54.—View across valley of rice fields, recently transplanted, in Kiangsi province, China.

levee construction. More than three hundred miles of sea
wall alone exist in the area covered by the sketch map, Fig.
52. The east bank of the Grand Canal, between Yang-
chow and Hwaianfu, is itself a great levee, holding back
the waters to the west above the eastern plain, diverting
them south, into the Yangtse kiang. But it is also provided
with spillways for use in times of excessive flood, permitting waters to discharge eastward. Such excess waters
however are controlled by another dike with canal along
its west side, some forty miles to the east, impounding the
water in a series of large lakes until it may gradually drain
away. This area is seen in Fig. 53, north of the Yangtse
river.

Along the banks of the Yangtse, and for many miles
along the Hwang ho, great levees have been built, sometimes in reenforcing series of two or three at different distances back from the channel where the stream bed is above
the adjacent country, in order to prevent widespread dis-
aster and to limit the inundated areas in times of unusual
flood. In the province of Hupeh, where the Han river
flows through two hundred miles of low country, this
stream is diked on both sides throughout the whole distance,
and in a portion of its course the hight of the levees reaches
thirty feet or more. Again, in the Canton delta region
there are other hundreds of miles of sea wall and dikes, so
that the aggregate mileage of this type of construction works
in the Empire can only be measured in thousands of miles.

In addition to the canal and levee construction works
there are numerous impounding reservoirs which are
brought into requisition to control overflow waters from the
great streams. Some of these reservoirs, like Tungting
lake in Hupeh and Poyang in Hunan, have areas of 2,000
and 1,800 square miles respectively and during the heaviest
rainy seasons each may rise through twenty to thirty feet,
Then there are other large and small lakes in the coastal
plain giving an aggregate reservoir area exceeding 13,000
square miles, all of which are brought into service in controlling flood waters, all of which are steadily filling with

Fig. 55.—Looking up the valley across terraced rice fields flooded with water, in Hunan province.

the sediments brought from the far away uncultivable mountain slopes and which are ultimately destined to become rich alluvial plains, doubtless to be canalized in the manner we have seen.

There is still another phase of these vast construction works which has been of the greatest moment in increasing the maintenance capacity of the Empire,—the wresting from the flood waters of the enormous volumes of silt which they carry, depositing it over the flooded areas, in the canals and along the shores in such manner as to add to the habitable and cultivable land. Reference has been made to the rapid growth of Chungming island in the mouth of the Yangtse kiang, and the million people now finding homes on the 270 square miles of newly made land which now has its canals, as may be seen in the upper margin of Fig. 52. The city of Shanghai, as its name signifies, stood originally on the seashore, which has now grown twenty miles to the northward and to the eastward. In 220 B. C. the town of Putai in Shantung stood one-third of a mile from the sea, but in 1730 it was forty-seven miles inland, and is forty-eight miles from the shore today.

Sienshuiku, on the Pei ho, stood upon the seashore in 500 A. D. We passed the city, on our way to Tientsin, eighteen miles inland. The dotted line laid in from the coast of the Gulf of Chihli in Fig. 53 marks one historic shore line and indicates a general growth of land eighteen miles to seaward.

Besides these actual extensions of the shore lines the centuries of flooding of lakes and low lying lands has so filled many depressions as to convert large areas of swamp into cultivated fields. Not only this, but the spreading of canal mud broadcast over the encircled fields has had two very important effects,—namely, raising the level of the low lying fields, giving them better drainage and so better physical condition, and adding new plant food in the form of virgin soil of the richest type, thus contributing to the maintenance of soil fertility, high maintenance capacity and permanent agriculture through all the centuries.

These operations of maintenance and improvement had a very early inception; they appear to have persisted throughout the recorded history of the Empire and are in vogue today. Canals of the type illustrated in Figs. 51 and 52 have been built between 1886 and 1901, both on the extensions of Chungming island and the newly formed main land to the north, as is shown by comparison of Stieler's atlas, revised in 1886, with the recent German survey.

Earlier than 2255 B. C., more than 4100 years ago, Emperor Yao appointed "The Great" Yu "Superintendent of Works" and entrusted him with the work of draining off the waters of disastrous floods and of canalizing the rivers, and he devoted thirteen years to this work. This great engineer is said to have written several treatises on agriculture and drainage, and was finally called, much against his wishes, to serve as Emperor during the last seven years of his life.

The history of the Hwang ho is one of disastrous floods and shiftings of its course, which have occurred many times in the years since before the time of the Great Yu, who perhaps began the works perpetuated today. Between 1300 A. D. and 1852 the Hwang ho emptied into the Yellow Sea south of the highlands of Shantung, but in that year, when in unusual flood, it broke through the north levees and finally took its present course, emptying again into the Gulf of Chihli, some three hundred miles further north. Some of these shiftings of course of the Hwang ho and of the Yangtse kiang are indicated in dotted lines on the sketch map, Fig. 53, where it may be seen that the Hwang ho during 146 years, poured its waters into the sea as far north as Tientsin, through the mouth of the Pei ho, four hundred miles to the northward of its mouth in 1852.

This mighty river is said to carry at low stage, past the city of Tsinan in Shantung, no less than 4,000 cubic yards of water per second, and three times this volume when running at flood. This is water sufficient to inundate thirty-three square miles of level country ten feet deep in twenty-four hours. What must be said of the mental status of a people who for forty centuries have measured their strength against such a Titan racing past their homes above the level of their fields, confined only between walls of their own construction? While they have not always succeeded in controlling the river, they have never failed to try again. In 1877 this river broke its banks, inundating a vast area, bringing death to a million people. Again, as late as 1898, fifteen hundred villages to the northeast of Tsinan and a much larger area to the southwest of the same city were devastated by it, and it is such events as these which have won for the river the names ''China's Sorrow,'' ''The Ungovernable'' and ''The Scourge of the Sons of Han.''

The building of the Grand Canal appears to have been a comparatively recent event in Chinese history. The middle section, between the Yangtse and Tsingkiangpu, is said to have been constructed about the sixth century B. C.; the southern section, between Chingkiang and Hangchow, during the years 605 to 617 A. D.; but the northern section, from the channel of the Hwang ho deserted in 1852, to Tientsin, was not built until the years 1280–1283.

While this canal has been called by the Chinese Yu ho (Imperial river), Yun ho (Transport river) or Yunliang ho (Tribute bearing river) and while it has connected the great rivers coming down from the far interior into a great water-transport system, this feature of construction may have been but a by-product of the great dominating purpose which led to the vast internal improvements in the form of canals, dikes, levees and impounding reservoirs so widely scattered, so fully developed and so effectively utilized. Rather the master purpose must have been maintenance for the increasing flood of humanity. And I am willing to grant to the Great Yu, with his finger on the pulse of the nation, the power to project his vision four thousand years into the future of his race and to formulate some of the measures which might be inaugurated to grow with the years and make certain perpetual maintenance for those to follow.

The exhaustion of cultivated fields must always have been the most fundamental, vital and difficult problem of all civilized people and it appears clear that such canalization as is illustrated in Figs. 51 and 52 may have been primarily initial steps in the reclamation of delta and overflow lands. At any rate, whether deliberately so planned or not, the canalization of the delta and overflow plains of China has been one of the most fundamental and fruitful measures for the conservation of her national resources that they could have taken, for we are convinced that this oldest nation in the world has thus greatly augmented the extension of its coastal plains, conserving and building out of the waste of erosion wrested from the great streams, hundreds of square miles of the richest and most enduring of soils, and we have little doubt that were a full and accurate account given of human influence upon the changes in this remarkable region during the last four thousand years it would show that these gigantic systems of canalization have been matters of slow, gradual growth, often initiated and always profoundly influenced by the labors of the strong, patient, persevering, thoughtful but ever silent husbandmen in their efforts to acquire homes and to maintain the productive power of their fields.

Nothing appears more clear than that the greatest material problem which can engage the best thought of China today is that of perfecting, extending and perpetuating the means for controlling her flood waters, for better draining of her vast areas of low land, and for utilizing the tremendous loads of silt borne by her streams more effectively in fertilizing existing fields and in building and reclaiming new land. With her millions of people needing homes and anxious for work; who have done so much in land building, in reclamation and in the maintenance of soil fertility, the government should give serious thought to the possibility of putting large numbers of them at work, effectively directed by the best engineering skill. It must now be entirely practicable, with engineering skill and mechanical appliances, to put the Hwang ho, and other rivers of China subject to overflow, completely under control. With the Hwang ho confined to its channel, the adjacent low lands can be better drained by canalization and freed from the accumulating saline deposits which are rendering them sterile. Warping may be resorted to during the flood season to raise the level of adjacent low-lying fields, rendering them at the same time more fertile. Where the river is running above the adjacent plains there is no difficulty in drawing off the turbid water by gravity, under controlled conditions, into diked basins, and even in compelling the river to buttress its own levees. There is certainly great need and great opportunity for China to make still better and more efficient her already wonderful transportation canals and those devoted to drainage, irrigation and fertilization.

In the United States, along the same lines, now that we are considering the development of inland waterways, the subject should be surveyed broadly and much careful study may well be given to the works these old people have developed and found serviceable through so many centuries. The Mississippi is annually bearing to the sea nearly 225,000 acre-feet of the most fertile sediment, and between levees along a raised bed through two hundred miles of country subject to inundation. The time is here when there should be undertaken a systematic diversion of a large part of this fertile soil over the swamp areas, building them into well drained, cultivable, fertile fields provided with waterways to serve for drainage, irrigation, fertilization and transportation. These great areas of swamp land may thus be converted into the most productive rice and sugar plantations to be found anywhere in the world, and the area made capable of maintaining many millions of people as long as the Mississippi endures, bearing its burden of fertile sediment.

But the conservation and utilization of the wastes of soil erosion, as applied in the delta plain of China, stupendous as this work has been, is nevertheless small when measured by the savings which accrue from the careful

Fig. 56.—Fields graded for the better conservation of rainfall and fertility and for more efficient irrigation and drainage, Chekiang province, China.

and extensive fitting of fields so largely practiced, which both lessens soil erosion and permits a large amount of soluble and suspended matter in the run-off to be applied to, and retained upon, the fields through their extensive systems of irrigation. Mountainous and hilly as are the lands of Japan, 11,000 square miles of her cultivated fields in the main islands of Honshu, Kyushu and Shikoku have been carefully graded to water level areas bounded by narrow raised rims upon which sixteen or more inches of run-off water, with its suspended and soluble matters, may be applied, a large part of which is retained on the fields or utilized by the crop, while surface erosion is almost completely prevented. The illustrations, Figs. 11, 12 and 13 show the application of the principle to the larger and more level fields, and in Figs. 151, 152 and 225 may be seen the practice on steep slopes.

If the total area of fields graded practically to a water level in Japan aggregates 11,000 square miles, the total area thus surface fitted in China must be eight or tenfold this amount. Such enormous field erosion as is tolerated at the present time in our southern and south Atlantic states is permitted nowhere in the Far East, so far as we observed, not even where the topography is much steeper. The tea orchards as we saw them on the steeper slopes, not level-terraced, are often heavily mulched with straw which makes erosion, even by heavy rains, impossible, while the treatment retains the rain where it falls, giving the soil opportunity to receive it under the impulse of both capillarity and gravity, and with it the soluble ash ingredients leached from the straw. The straw mulches we saw used in this manner were often six to eight inches deep, thus constituting a dressing of not less than six tons per acre, carrying 140 pounds of soluble potassium and 12 pounds of phosphorus. The practice, therefore, gives at once a good fertilizing, the highest conservation and utilization of rainfall, and a complete protection against soil erosion. It is a *multum in parvo* treatment which characterizes so many of the practices of these people, which have crystalized from twenty centuries of high tension experience.

In the Kiangsu and Chekiang provinces, as elsewhere in the densely populated portions of the Far East, we found almost all of the cultivated fields very nearly level or made so by grading. Instances showing the type of this grading in a comparatively level country are seen in Figs. 56 and 57. By this preliminary surface fitting of the fields these people have reduced to the lowest possible limit the waste of soil fertility by erosion and surface leaching. At the same time they are able to retain upon the field, uniformly distributed over it, the largest part of the rainfall practicable, and to compel a much larger proportion of the necessary run off to leave by under-drainage than would be possible otherwise, conveying the plant food developed in the surface soil to the roots of the crops, while they make possible a more complete absorption and retention by the soil of the soluble plant food materials not taken up. This same treatment also furnishes the best possible conditions for the application of water to the fields when supplemental irrigation would be helpful, and for the withdrawal of surplus rainfall by surface drainage, should this be necessary.

Besides this surface fitting of fields there is a wide application of additional methods aiming to conserve both rainfall and soil fertility, one of which is illustrated in Fig. 58, showing one end of a collecting reservoir. There were three of these reservoirs in tandem, connected with each other by surface ditches and with an adjoining canal. About the reservoir the level field is seen to be thrown into beds with shallow furrows between the long narrow ridges. The furrows are connected by a head drain around the margin of the reservoir and separated from it by a narrow raised rim. Such a reservoir may be six to ten feet deep but can be completely drained only by pumping or by evaporation during the dry season. Into such reservoirs the excess surface water is drained where all suspended matter carried from the field collects and is returned, either

8

Fig. 57.—Illustrating the same practice as Fig. 56. The crops in both instances are windsor beans and rape on the low-lying areas, and mulberries on the higher ground. Rice will follow on the lower areas.

Fig. 58.—Collecting reservoir for the conservation of rainfall and fertility, used also as fish ponds and to provide water and mud for making composts. The circular ring in the foreground is a compost pit.

directly as an application of mud or as material used in composts. In the preparation of composts, pits are dug near the margin of the reservoir, as seen in the illustration, and into them are thrown coarse manure and any roughage in the form of stubble or other refuse which may be available, these materials being saturated with the soft mud dipped from the bottom of the reservoir.

Fig. 59.—Two compost pits filled with roughage and mud from the canal, in preparation of compost for the fields. The narrow path along the canal is one of the common thoroughfares in Kiangsu province.

In all of the provinces where canals are abundant they also serve as reservoirs for collecting surface washings and along their banks great numbers of compost pits are maintained and repeatedly filled during the season, for use on the fields as the crops are changed. Fig. 59 shows two such pits on the bank of a canal, already filled.

In other cases, as in the Shantung province, illustrated in Fig. 60, the surface of the field may be thrown into broad leveled lands separated and bounded by deep and wide trenches into which the excess water of very heavy rains may collect. As we saw them there was no provision for draining the trenches and the water thus collected either seeps away or evaporates, or it may be returned in part by underflow and capillary rise to the soil from which it was collected, or be applied directly for irrigation

Fig. 60.—Trenching of fields for drainage, conservation of rainfall and of fertility, in the Shantung province. Trenches are two feet wide on the bottom, six or eight feet wide at the top, and two and a half to three feet deep.

by pumping. In this province the rains may often be heavy but the total fall for the year is small, being little more than twenty-four inches, hence there is the greatest need for its conservation, and this is carefully practiced.

VI.

SOME CUSTOMS OF THE COMMON PEOPLE.

The Tosa Maru brought us again into Shanghai March 20th, just in time for the first letters from home. A ricksha man carried us and our heavy valise at a smart trot from the dock to the Astor House, more than a mile, for 8.6 cents, U. S. currency, and more than the conventional price for the service rendered. On our way we passed several loaded carryalls of the type seen in Fig. 61, on which women were riding for a fare one-tenth that we had paid, but at a slower pace and with many a jolt.

The ringing chorus which came loud and clear when yet half a block away announced that the pile drivers were still at work on the foundation for an annex to the Astor House, and so were they on May 27th when we returned from the Shantung province, 88 days after we saw them first, but with the task then practically completed. Had the eighteen men labored continuously through this interval, the cost of their services to the contractor would have been but $205.92. With these conditions the engine-driven pile driver could not compete. All ordinary labor here receives a low wage. In the Chekiang province farm labor employed by the year received $30 and board, ten years ago, but now is receiving $50. This is at the rate of about $12.90 and $21.50, gold, materially less than there is paid per month in the United States. At Tsingtao in the Shantung province a missionary was paying a Chinese cook ten dollars per month, a man for general work nine dollars per month, and the cook's wife, for doing the mending and other family service, two dollars per month, all living at home and feeding themselves. This service rendered for $9.03, gold, per month covers the marketing, all care of the garden and lawn as well as all the work in the

Fig. 61.—A common means of transport on the streets of Shanghai, used much more frequently by women than by men.

house. Missionaries in China find such servants reliable and satisfactory, and trust them with the purse and the marketing for the table, finding them not only honest but far better at a bargain and at economical selection than themselves.

We had a soil tube made in the shops of a large English ship building and repair firm, employing many hundred Chinese as mechanics, using the most modern and complex machinery, and the foreman stated that as soon as the men could understand well enough to take orders they were even better shop hands than the average in Scotland and England. An educated Chinese booking clerk at the Soochow railway station in Kiangsu province was receiving a salary of $10.75, gold, per month. We had inquired the way to the Elizabeth Blake hospital and he volunteered to escort us and did so, the distance being over a mile.

Fig. 62.—A sewing circle in the open air and sunshine, Shanghai.

He would accept no compensation, and yet I was an entire stranger, without introduction of any kind.

Everywhere we went in China, the laboring people appeared generally happy and contented if they have something to do, and showed clearly that they were well nourished. The industrial classes are thoroughly organized, having had their guilds or labor unions for centuries and it is not at all uncommon for a laborer who is known to

have violated the rules of his guild to be summarily dealt
with or even to disappear without questions being asked.
In going among the people, away from the lines of tourist
travel, one gets the impression that everybody is busy or
is in the harness ready to be busy. Tramps of our hobo
type have few opportunities here and we doubt if one
exists in either of these countries. There are people physically disabled who are asking alms and there are organized
charities to help them, but in proportion to the total
population these appear to be fewer than in America or
Europe. The gathering of unfortunates and habitual
beggars about public places frequented by people of leisure
and means naturally leads tourists to a wrong judgment
regarding the extent of these social conditions. Nowhere
among these densely crowded people, either Chinese, Japanese or Korean, did we see one intoxicated, but among
Americans and Europeans many instances were observed.
All classes and both sexes use tobacco and the British-
American Tobacco Company does a business in China
amounting to millions of dollars annually.

During five months among these people we saw but two
children in a quarrel. The two little boys were having
their trouble on Nanking road, Shanghai, where, grasping
each other's pigtails, they tussled with a vengeance until
the mother of one came and parted their ways.

Among the most frequent sights in the city streets
are the itinerant venders of hot foods and confections.
Stove, fuel, supplies and appliances may all be carried
on the shoulders, swinging from a bamboo pole. The
mother in Fig. 63 was quite likely thus supporting her
family and the children are seen at lunch, dressed in
the blue and white calico prints so generally worn by
the young. The printing of this calico by the very ancient, simple yet effective method we witnessed in the
farm village along the canal seen in Fig. 10. This
art, as with so many others in China, was the inheritance
of the family we saw at work, handed down to them
through many generations. The printer was standing

at a rough work bench upon which a large heavy stone
in cubical form served as a weight to hold in place
a thoroughly lacquered sheet of tough cardboard in which
was cut the pattern to appear in white on the cloth.
Beside the stone stood a pot of thick paste prepared from
a mixture of lime and soy bean flour.   The soy beans

Fig. 63.—Eating lunch.

were being ground in one corner of the same room by a
diminutive edition of such an outfit as seen in Fig. 64.
The donkey was working in his permanent abode and
whenever off duty he halted before manger and feed.   At
the operator's right lay a bolt of white cotton cloth fixed
to unroll and pass under the stencil, held stationary by
the heavy weight.   To print, the stencil was raised and the
cloth brought to place under it.   The paste was then deftly

spread with a paddle over the surface and thus upon the cloth beneath wherever exposed through the openings in the stencil. This completes the printing of the pattern on one section of the bolt of cloth. The free end of the stencil is then raised, the cloth passed along the proper distance by hand and the stencil dropped in place for the next application. The paste is permitted to dry upon the cloth and when the bolt has been dipped into the blue

Fig. 64.—Stone mill in common use for grinding beans and various kinds of grain.

dye the portions protected by the paste remain white. In this simple manner has the printing of calico been done for centuries for the garments of millions of children. From the ceiling of the drying room in this printery of olden times were hanging some hundreds of stencils bearing different patterns. In our great calico mills, printing hundreds of yards per minute, the mechanics and the chemistry differ only in detail of application and in dispatch, not in fundamental principle.

In almost any direction we travelled outside the city, in the pleasant mornings when the air was still, the laying of warp for cotton cloth could be seen, to be woven later in the country homes. We saw this work in progress many times and in many places in the early morning, usually along some roadside or open place, as seen in Fig. 65, but never later in the day. When the warp is laid each will be rolled upon its stretcher and removed to the house to be woven.

In many places in Kiangsu province batteries of the large dye pits were seen sunk in the fields and lined with

Fig. 65.—Laying warp in the country for four bolts of cotton cloth.

cement. These were six to eight feet in diameter and four to five feet deep. In one case observed there were nine pits in the set. Some of the pits were neatly sheltered beneath live arbors, as represented in Fig. 66. But much of this spinning, weaving, dyeing and printing of late years is being displaced by the cheaper calicos of foreign make and most of the dye pits we saw were not now used for this purpose, the two in the illustration serving as manure receptacles. Our interpreter stated however that there is a growing dissatisfaction with foreign goods on account of their lack of durability; and we saw many cases where the cloth dyed blue was being dried in large quantities on the grave lands.

In another home for nearly an hour we observed a method of beating cotton and of laying it to serve as the body for mattresses and the coverlets for beds. This we could do without intrusion because the home was also the work shop and opened full width directly upon the narrow street. The heavy wooden shutters which closed the home at night were serving as a work bench about seven feet square, laid upon movable supports. There was barely

Fig. 66.—Two dye pits under woven arbor shelter, now abandoned for their original purpose and used as manure receptacles. The trees in the rear are a typical clump of bamboo so frequently seen about farm houses.

room to work between it and the sidewalk without impeding traffic, and on the three other sides there was a floor space three or four feet wide. In the rear sat grandmother and wife while in and out the four younger children were playing. Occupying the two sides of the room were receptacles filled with raw cotton and appliances for the work. There may have been a kitchen and sleeping room behind but no door, as such, was visible. The finished mattresses, carefully rolled and wrapped in paper, were suspended from the ceiling. On the improvised work table, with its top two feet above the floor, there had been laid in the morning before our visit, a mass of soft white cotton more than six feet square and fully twelve inches deep. On opposite sides of this table the father and his son, of twelve years, each twanged the string of their heavy bamboo bows, snapping the lint from the wads of cotton and flinging it broadcast in an even layer over the surface of the growing mattress, the two strings the while emitting tones pitched far below the hum of the bumblebee. The heavy bow was steadied by a cord secured around the

Fig 67.—Japanese form of bow used in the home for spreading cotton in making wadding and cotton batting.

body of the operator, allowing him to manage it with one hand and to move readily around his work in a manner different from the custom of the Japanese seen in Fig. 67. By this means the lint was expeditiously plucked and skillfully and uniformly laid, the twanging being effected by an appliance similar to that used in Japan.

Repeatedly, taken in small bits from the barrel of cotton, the lint was distributed over the entire surface with great dexterity and uniformity, the mattress growing upward with perfectly vertical sides, straight edges and square corners. In this manner a thoroughly uniform texture is secured which compresses into a body of even thickness, free from hard places.

The next step in building the mattress is even more simple and expeditious. A basket of long bobbins of roughly spun cotton was near the grandmother and probably her handiwork. The father took from the wall a slender bamboo rod like a fish-pole, six feet long, and selecting one of the spools, threaded the strand through an eye in the small end. With the pole and spool in one hand and the free end of the thread, passing through the eye, in the other, the father reached the thread across the mattress to the boy who hooked his finger over it, carrying it to one edge of the bed of cotton. While this was doing the father had whipped the pole back to his side and caught the thread over his own finger, bringing this down upon the cotton opposite his son. There was thus laid a double strand, but the pole continued whipping back and forth across the bed, father and son catching the threads and bringing them to place on the cotton at the rate of forty to fifty courses per minute, and in a very short time the entire surface of the mattress had been laid with double strands. A heavy bamboo roller was next laid across the strands at the middle, passed carefully to one side, back again to the middle and then to the other edge. Another layer of threads was then laid diagonally and this similarly pressed with the same roller; then another diagonally the other way and finally straight across in both directions. A similar network of strands had been laid upon the table before spreading the cotton. Next a flat bottomed, circular, shallow basketlike form two feet in diameter was used to gently compress the material from twelve to six inches in thickness. The woven threads were now turned over the edge of the mattress on all sides and sewed down, after which, by means of two heavy solid wooden disks eighteen inches in diameter, father and son compressed the cotton until the thickness was reduced to three inches. There remained the task of carefully folding and wrapping the finished piece in oiled paper and of suspending it from the ceiling.

On March 20th, when visiting the Boone Road and Nanking Road markets in Shanghai, we had our first surprise regarding the extent to which vegetables enter into the daily diet of the Chinese. We had observed long processions of wheelbarrow men moving from the canals through the streets carrying large loads of the green tips of rape in bundles a foot long and five inches in diameter. These had come from the country on boats each carrying tons of the succulent leaves and stems. We had counted as many as fifty wheelbarrow men passing a given point on the street in quick succession, each carrying 300 to 500 pounds of the green rape and moving so rapidly that it was not easy to keep pace with them, as we learned in following one of the trains during twenty minutes to its destination. During this time not a man in the train halted or slackened his pace.

This rape is very extensively grown in the fields, the tips of the stems cut when tender and eaten, after being boiled or steamed, after the manner of cabbage. Very large quantities are also packed with salt in the proportion of about twenty pounds of salt to one hundred pounds of the rape. This, Fig. 68, and many other vegetables are sold thus pickled and used as relishes with rice, which invariably is cooked and served without salt or other seasoning.

Another field crop very extensively grown for human food, and partly as a source of soil nitrogen, is closely allied to our alfalfa. This is the *Medicago astragalus,* two beds of which are seen in Fig. 69. Tender tips of the stems are gathered before the stage of blossoming is reached and served as food after boiling or steaming. It is known among the foreigners as Chinese "clover." The stems are also cooked and then dried for use when the crop is out of season. When picked *very* young, wealthy Chinese families pay an extra high price for the tender shoots, sometimes as much as 20 to 28 cents, our currency, per pound.

Fig. 68.—"Salted cabbage," prepared from young rape, displayed for sale in Boone Road market, Shanghai.

The markets are thronged with people making their purchases in the early mornings, and the congested condition, with the great variety of vegetables, makes it almost as impressive a sight as Billingsgate fish market in London. In the following table we give a list of vegetables observed there and the prices at which they were selling.

LIST OF VEGETABLES DISPLAYED FOR SALE IN BOONE ROAD MARKET, SHANGHAI, APRIL 6TH, 1909, WITH PRICES EXPRESSED IN U. S. CURRENCY.

	Cents.		Cents.
Lotus roots, per lb	1.60	Tee Tsai, per lb	.53
Bamboo sprouts, per lb	6.40	Chinese celery, per lb	.67
English cabbage, per lb	1.33	Chinese clover, per lb	.53
Olive greens, per lb	.67	Chinese clover, very young, lb	21.33
White greens, per lb	.33	Oblong white cabbage, per lb	2.00

	Cents.
Red beans, per lb	1.33
Yellow beans, per lb	1.87
Peanuts, per lb	2.49
Ground nuts, per lb	2.96
Cucumbers, per lb	2.58
Green pumpkin, per lb	1.62
Maize, shelled, per lb	1.00
Windsor beans, dry, per lb	1.72
French lettuce, per head	.44
Hau Tsai, per head	.87
Cabbage lettuce, per head	.22
Kale, per lb	1.60
Rape, per lb	.23
Portuguese water cress, basket	2.15
Shang tsor, basket	8.60
Carrots, per lb	.97
String beans, per lb	1.60
Irish potatoes, per lb	1.60
Red onions, per lb	4.96
Long white turnips, per lb	.44
Flat string beans, per lb	4.80
Small white turnips, bunch	.44
Onion stems, per lb	1.29
Lima beans, green, shelled, lb	6.45
Egg plants, per lb	4.30
Tomatoes, per lb	5.16
Small flat turnips, per lb	.86
Small red beets, per lb	1.29
Artichokes, per lb	1.29

	Cents.
White beans, dry, per lb	4.30
Radishes, per lb	1.29
Garlic, per lb	2.15
Kohl rabi, per lb	2.15
Mint, per lb	4.30
Leeks, per lb	2.13
Large celery, bleached, bunch	2.10
Sprouted peas, per lb	.80
Sprouted beans, per lb	.93
Parsnips, per lb	1.29
Ginger roots, per lb	1.60
Water chestnuts, per lb	1.33
Large sweet potatoes, per lb	1.33
Small sweet potatoes, per lb	1.00
Onion sprouts, per lb	2.13
Spinach, per lb	1.00
Fleshy stemmed lettuce, peeled, per lb	2.00
Fleshy stemmed lettuce, unpeeled, per lb	.67
Bean curd, per lb	3.93
Shantung walnuts, per lb	4.30
Duck eggs, dozen	8.34
Hen's eggs, dozen	7.30
Goat's meat, per lb	6.45
Pork, per lb	6.88
Hens, live weight, per lb	6.45
Ducks, live weight, per lb	5.59
Cockerels, live weight, per lb	5.59

This long list, made up chiefly of fresh vegetables displayed for sale on one market day, is by no means complete. The record is only such as was made in passing down one side and across one end of the market occupying nearly one city block. Nearly everything is sold by weight and the problem of correct weights is effectively solved by each purchaser carrying his own scales, which he unhesitatingly uses in the presence of the dealer. These scales are made on the pattern of the old time steelyards but from slender rods of wood or bamboo provided with a scale and sliding poise, the suspensions all being made with strings.

We stood by through the purchasing of two cockerels and the dickering over their weight. A dozen live birds were under cover in a large, open-work basket. The customer took out the birds one by one, examining them by touch, finally selecting two, the price being named. These the dealer tied together by their feet and weighed them, announcing the result; whereupon the customer checked the statement with his own scales. An animated dialogue followed, punctuated with many gesticulations and with the customer tossing the birds into the basket and turning to go away while the dealer grew more earnest. The purchaser finally turned back, and again balancing the roosters upon his scales, called a bystander to read the weight, and then flung them in apparent disdain at the dealer, who caught them and placed them in the customer's

Fig 69.—Two beds of Chinese clover (*Medicago astragalus*) grown in the garden for human food in the season and for soil fertility later.

basket. The storm subsided and the dealer accepted 92c, Mexican, for the two birds. They were good sized roosters and must have dressed more than three pounds each, yet for the two he paid less than 40 cents in our currency.

Bamboo sprouts are very generally used in China, Korea and Japan and when one sees them growing they suggest giant stalks of asparagus, some of them being three and even five inches in diameter and a foot in hight at the stage for cutting. They are shipped in large quantities from province to province where they do not grow or when they are out of season.   Those we saw in Nagasaki, referred to in Fig. 22, had come from Canton or Swatow

Fig. 70.—Boone Road vegetable market, April 6th, Shanghai, China.   The large vegetables in the lower section are lotus roots.

or possibly Formosa. The form, foliage and bloom of the bamboo give the most beautiful effects in the landscape, especially when grouped with tree forms. They are usually cultivated in small clumps about dwellings in places not otherwise readily utilized, as seen in Fig. 66. Like the asparagus bud, the bamboo sprout grows to its full hight between April and August, even when it exceeds thirty or even sixty feet in hight. The buds spring from fleshy underground stems or roots whose stored nourishment permits this rapid growth, which in its earlier stages may

Fig. 71.—Lotus pond with plant in bloom; cultivated for their fleshy roots used for food, shown in Fig. 70.

exceed twelve inches in twenty-four hours. But while the full size of the plant is attained the first season, three or four years are required to ripen and harden the wood sufficiently to make it suitable for the many uses to which the stems are put. It would seem that the time must come when some of the many forms of bamboo will be introduced and largely grown in many parts of this country.

Lotus roots form another article of diet largely used and widely cultivated from Canton to Tokyo. These are seen in the lower section of Fig. 70, and the plants in bloom in Fig. 71, growing in water, their natural habitat. The lotus is grown in permanent ponds not readily drained for rice or other crops, and the roots are widely shipped.

Sprouted beans and peas of many kinds and the sprouts of other vegetables, such as onions, are very generally seen in the markets of both China and Japan, at least during the late winter and early spring, and are sold as foods, having different flavors and digestive qualities, and no doubt with important advantageous effects in nutrition.

Ginger is another crop which is very widely and extensively cultivated. It is generally displayed in the market in the root form. No one thing was more generally hawked about the streets of China than the water chestnut. This is a small corm or fleshy bulb having the shape and size of a small onion. Boys pare them and sell a dozen spitted together on slender sticks the length of a knitting needle. Then there are the water caltropes, grown in the canals, producing a fruit resembling a horny nut having a shape which suggests for them the name "buffalo-horn". Still another plant, known as water-grass *(Hydropyrum latifolium)* is grown in Kiangsu province where the land is too wet for rice. The plant has a tender succulent crown of leaves and the peeling of the outer coarser ones away suggests the husking of an ear of green corn. The portion eaten is the central tender new growth, and when cooked forms a delicate savory dish. The farmers' selling price is three to four dollars, Mexican, per hundred catty, or $.97 to $1.29 per hundredweight, and the return per acre is from $13 to $20.

The small number of animal products which are included in the market list given should not be taken as indicating the proportion of animal to vegetable foods in the dietaries of these people. It is nevertheless true that they are vegetarians to a far higher degree than are most western nations, and the high maintenance efficiency of the agriculture of China, Korea and Japan is in great measure rendered possible by the adoption of a diet so largely vegetarian. Hopkins, in his Soil Fertility and Permanent Agriculture, page 234, makes this pointed statement of fact: "1000 bushels of grain has at least five times as much food value and will support five times as many people as will the meat or milk that can be made from it". He also calls attention to the results of many Rothamsted feeding experiments with growing and fattening cattle, sheep and swine, showing that the cattle destroyed outright, in every 100 pounds of dry substance eaten, 57.3 pounds, this passing off into the air, as does all of wood except the ashes, when burned in the stove; they left in the excrements 36.5 pounds, and stored as increase but 6.2 pounds of the 100. With sheep the corresponding figures were 60.1 pounds; 31.9 pounds and 8 pounds; and with swine they were 65.7 pounds; 16.7 pounds and 17.6 pounds. But less than two-thirds of the substance stored in the animal can become food for man and hence we get but four pounds in one hundred of the dry substances eaten by cattle in the form of human food; but five pounds from the sheep and eleven pounds from swine.

In view of these relations, only recently established as scientific facts by rigid research, it is remarkable that these very ancient people came long ago to discard cattle as milk and meat producers; to use sheep more for their pelts and wool than for food; while swine are the one kind of the three classes which they did retain in the role of middleman as transformers of coarse substances into human food.

It is clear that in the adoption of the succulent forms of vegetables as human food important advantages are gained. At this stage of maturity they have a higher digestibility, thus making the elimination of the animal less difficult. Their nitrogen content is relatively higher and this in a measure compensates for loss of meat. By devoting the soil to growing vegetation which man can directly digest they have saved 60 pounds per 100 of absolute waste by the animal, returning their own wastes to the field for the maintenance of fertility. In using these immature forms of vegetation so largely as food they are able to produce an immense amount that would otherwise be impossible, for this is grown in a shorter time, permitting the same soil to produce more crops. It is also produced late in the fall and early in the spring when the season is too cold and the hours of sunshine too few each day to permit of ripening crops.

VII.

THE FUEL PROBLEM, BUILDING AND TEXTILE MATERIALS.

With the vast and ever increasing demands made upon materials which are the products of cultivated fields, for food, for apparel, for furnishings and for cordage, better soil management must grow more important as populations multiply. With the increasing cost and ultimate exhaustion of mineral fuel; with our timber vanishing rapidly before the ever growing demands for lumber and paper; with the inevitably slow growth of trees and the very limited areas which the world can ever afford to devote to forestry, the time must surely come when, in short period rotations, there will be grown upon the farm materials from which to manufacture not only paper and the substitutes for lumber, but fuels as well. The complete utilization of every stream which reaches the sea, reinforced by the force of the winds and the energy of the waves which may be transformed along the coast lines, cannot fully meet the demands of the future for power and heat; hence only in the event of science and engineering skill becoming able to devise means for transforming the unlimited energy of space through which we are ever whirled, with an economy approximating that which crops now exhibit, can good soil management be relieved of the task of meeting a portion of the world's demand for power and heat.

When these statements were made in 1905 we did not know that for centuries there had existed in China, Korea

and Japan a density of population such as to require the
extensive cultivation of crops for fuel and building material, as well as for fabrics, by the ordinary methods of
tillage, and hence another of the many surprises we had
was the solution these people had reached of their fuel
problem and of how to keep warm. Their solution has
been direct and the simplest possible. Dress to make fuel
for warmth of body unnecessary, and burn the coarser
stems of crops, such as cannot be eaten, fed to animals
or otherwise made useful. These people still use what
wood can be grown on the untillable land within transporting distance, and convert much wood into charcoal, making
transportation over longer distances easier. The general
use of mineral fuels, such as coal, coke, oils and gas, had
been impossible to these as to every other people until
within the last one hundred years. Coal, coke, oil and
natural gas, however, have been locally used by the Chinese
from very ancient times. For more than two thousand
years brine from many deep wells in Szechwan province
has been evaporated with heat generated by the burning
of natural gas from wells, conveyed through bamboo stems
to the pans and burned from iron terminals. In other
sections of the same province much brine is evaporated
over coal fires. Alexander Hosie estimates the production
of salt in Szechwan province at more than 600 million
pounds annually.

Coal is here used also to some extent for warming the
houses, burned in pits sunk in the floor, the smoke escap-
ing where it may. The same method of heating we saw
in use in the post office at Yokohama during February.
The fires were in large iron braziers more than two feet
across the top, simply set about the room, three being in
operation. Stoves for house warming are not used in
dwellings in these countries.

In both China and Japan we saw coal dust put into
the form and size of medium oranges by mixing it with
a thin paste of clay. Charcoal is similarly molded, as

seen in Fig. 72, using a byproduct from the manufacture of rice syrup for cementing. In Nanking we watched with much interest the manufacture of charcoal briquets by another method. A Chinese workman was seated upon the earth floor of a shop. By his side was a pile of powdered charcoal, a dish of rice syrup byproduct and a basin of the moistened charcoal powder. Between his legs was a heavy mass of iron containing a slightly conical mold two inches deep, two and a half inches across at the top and a heavy iron hammer weighing several pounds. In his left hand he held a short heavy ramming tool and with his right placed in the mold a pinch of the moistened

Fig. 72.—Charcoal balls briquetted with rice water or clay, for use as fuel.

charcoal; then followed three well directed blows from the hammer upon the ramming tool, compressing the charge of moistened, sticky charcoal into a very compact layer. Another pinch of charcoal was added and the process repeated until the mold was filled, when the briquet was forced out.

By this simplest possible mechanism, the man, utilizing but a small part of his available energy, was subjecting the charcoal to an enormous pressure such as we attain only with the best hydraulic presses, and he was using the principle of repeated small charges recently patented and applied in our large and most efficient cotton and hay presses, which permit much denser bales to be made than is possible when large charges are added, and the Chinese is here, as in a thousand other ways, thoroughly sound in his application of mechanical principles. His output for the day was small but his patience seemed unlimited. His arms and body, bared to the waist, showed vigor and good feeding, while his face wore the look of contentment.

With forty centuries of such inheritance coursing in the veins of four hundred millions of people, in a country possessed of such marvelous wealth of coal and water power, of forest and of agricultural possibilities, there should be a future speedily blossoming and ripening into all that is highest and best for such a nation. If they will retain their economies and their industry and use their energies to develop, direct and utilize the power in their streams and in their coal fields along the lines which science has now made possible to them, at the same time walking in paths of peace and virtue, there is little worth while which may not come to such a people.

A Shantung farmer in winter dress, Fig. 18, and the Kiangsu woman portrayed in Fig. 73, in corresponding costume, are typical illustrations of the manner in which food for body warmth is minimized and of the way the heat generated in the body is conserved. Observe his wadded and quilted frock, his trousers of similar goods tied about the ankle, with his feet clad in multiple socks and cloth shoes provided with thick felted soles. These types of dress, with the wadding, quilting, belting and tying, incorporate and confine as part of the effective material a large volume of air, thus securing without cost, much additional warmth without increasing the weight of the garments. Beneath these outer garments several under pieces of different weights are worn which greatly conserve the warmth during the coldest weather and make possible a wide range of adjustment to suit varying changes in temperature. It is doubtful if there could be devised a wardrobe suited to the conditions of these people at a smaller first cost and maintenance expense. Rev. E. A. Evans, of the China Inland Mission, for many years residing at Sunking in Szechwan, estimated that a farmer's wardrobe, once it was procured, could be maintained with an annual expenditure of $2.25 of our currency, this sum procuring the materials for both repairs and renewals.

The intense individual economy, extending to the smallest matters, so universally practiced by these people, has sustained the massive strength of the Mongolian nations

Fig. 73.—A Kiangsu country woman in winter dress.

through their long history and this trait is seen in their handling of the fuel problem, as it is in all other lines. In the home of Mrs. Wu, owner and manager of a 25-acre rice farm in Chekiang province, there was a masonry kang seven by seven feet, about twenty-eight inches high, which could be warmed in winter by building a fire within. The top was fitted for mats to serve as couch by day and as a place upon which to spread the bed at night. In the Shantung province we visited the home of a prosperous farmer and here found two kangs in separate sleeping apartments, both warmed by the waste heat from the kitchen whose chimney flue passed horizontally under the kangs before rising through the roof. These kangs were wide enough to spread the beds upon, about thirty inches high, and had been constructed from brick twelve inches square and four inches thick, made from the clay subsoil taken from the fields and worked into a plastic mass, mixed with chaff and short straw, dried in the sun and then laid in a mortar of the same material. These massive kangs are thus capable of absorbing large amounts of the waste heat from the kitchen during the day and of imparting congenial warmth to the couches by day and to the beds and sleeping apartments during the night. In some Manchurian inns large compound kangs are so arranged that the guests sleep heads together in double rows, separated only by low dividing rails, securing the greatest economy of fuel, providing the guests with places where they may sit upon the moderately warmed fireplace, and spread their beds when they retire.

The economy of the chimney beds does not end with the warmth conserved. The earth and straw brick, through the processes of fermentation and through shrinkage, become open and porous after three or four years of service, so that the draft is defective, giving annoyance from smoke, which requires their renewal. But the heat, the fermentation and the absorption of products of combustion have together transformed the comparatively infertile subsoil into what they regard as a valuable fertilizer and these discarded brick are used in the preparation of compost fertilizers for the fields. On account of this value of the discarded brick the large amount of labor involved in removing and rebuilding the kangs is not regarded altogether as labor lost.

Our own observations have shown that heating soils to dryness at a temperature of 110° C. greatly increases the freedom with which plant food may be recovered

from them by the solvent power of water, and the same
heating doubtless improves the physical and biological conditions of the soil as well. Nitrogen combined as ammonia,
and phosphorus, potash and lime are all carried with the
smoke or soot, mechanically in the draft and arrested upon
the inner walls of the kangs or filter into the porous brick
with the smoke, and thus add plant food directly to the
soil. Soot from wood has been found to contain, as an
average, 1.36 per cent of nitrogen; .51 per cent of phosphorus and 5.34 per cent of potassium. We practice burning straw and corn stalks in enormous quantities, to get
them easily out of the way, thus scattering on the winds
valuable plant food, thoughtlessly and lazily wasting
where these people laboriously and religiously save. These
are gains in addition to those which result from the formation of nitrates, soluble potash and other plant foods
through fermentation. We saw many instances where
these discarded brick were being used, both in Shantung
and Chihli provinces, and it was common in walking
through the streets of country villages to see piles of them,
evidently recently removed.

The fuel grown on the farms consists of the stems of
all agricultural crops which are to any extent woody,
unless they can be put to some better use. Rice straw,
cotton stems pulled by the roots after the seed has been
gathered, the stems of windsor beans, those of rape and
the millets, all pulled by the roots, and many other kinds,
are brought to the market tied in bundles in the manner
seen in Figs. 74, 75 and 76. These fuels are used for
domestic purposes and for the burning of lime, brick,
roofing tile and earthenware as well as in the manufacture
of oil, tea, bean-curd and many other processes. In the
home, when the meals are cooked with these light bulky
fuels, it is the duty of some one, often one of the children,
to sit on the floor and feed the fire with one hand while
with the other a bellows is worked to secure sufficient
draft.

Fig. 74.—Boat loads of fuel, mainly bundles of rice straw and cotton stems, on Soochow creek, Shanghai.

The manufacture of cotton seed oil and cotton seed cake is one of the common family industries in China, and in one of these homes we saw rice hulls and rice straw being used as fuel. In the large low, one-story, tile-roofed building serving as store, warehouse, factory and dwelling, a family of four generations were at work, the grandfather supervising in the mill and the grandmother

Fig. 75.—Cotton stem fuel being conveyed from the canals to city market stalls.

leading in the home and store where the cotton seed oil was being retailed for 22 cents per pound and the cotton seed cake at 33 cents, gold, per hundredweight. Back of the store and living rooms, in the mill compartment, three blindfolded water buffalo, each working a granite mill, were crushing and grinding the cotton seed. Three other buffalo, for relay service, were lying at rest or eating, awaiting their turn at the ten-hour working day. Two of the mills were horizontal granite burrs more than four

Fig. 76.—Rice straw fuel being conveyed from canal boats to city market stalls.

feet in diameter, the upper one revolving once with each circuit made by the cow. The third mill was a pair of massive granite rollers, each five feet in diameter and two feet thick, joined on a very short horizontal axle which revolved on a circular stone plate about a vertical axis once with each circuit of the buffalo. Two men tended the three mills. After the cotton seed had been twice passed through the mills it was steamed to render the

Fig. 77.—Appliance for steaming tea leaves, used in Japan and the same in principle as used in China for steaming meal from which oil is to be expressed.

oil fluid and more readily expressed. The steamer consisted of two covered wooden hoops not unlike that seen in Fig. 77, provided with screen bottoms, and in these the meal was placed over openings in the top of an iron kettle of boiling water from which the steam was forced through the charge of meal. Each charge was weighed in a scoop balanced on the arm of a bamboo scale, thus securing a uniform weight for the cakes.

On the ground in front of the furnace sat a boy of

twelve years steadily feeding rice chaff into the fire with his left hand at the rate of about thirty charges per minute, while with his right hand, and in perfect rhythm, he drew back and forth the long plunger of a rectangular box bellows, maintaining a forced draft for the fire. At intervals the man who was bringing fuel fed into the furnace a bundle of rice straw, thus giving the boy's left arm a moment's respite. When the steaming has rendered the oil sufficiently fluid the meal is transferred, hot, to ten-inch hoops two inches deep, made of braided bamboo strands, and is deftly tramped with the bare feet, while hot, the operator steadying himself by a pair of hand bars. After a stack of sixteen hoops, divided by a slight sifting of chaff or short straw to separate the cakes, had been completed these were taken to one of four pressmen, who were kept busy in expressing the oil.

The presses consisted of two parallel timbers framed together, long enough to receive the sixteen hoops on edge above a gap between them. These cheeses of meal are subjected to an enormous pressure secured by means of three parallel lines of wedges forced against the follower each by an iron-bound master wedge, driven home with a heavy beetle weighing some twenty-five or thirty pounds. The lines of wedges were tightened in succession, the loosened line receiving an additional wedge to take up the slack after drawing back the master wedge, which was then driven home. To keep good the supply of wedges which are often crushed under the pressure a second boy, older than the one at the furnace, was working on the floor, shaping new ones, the broken wedges and the chips going to the furnace for fuel.

By this very simple, readily constructed and inexpensive mechanism enormous pressures were secured and when the operator had obtained the desired compression he lighted his pipe and sat down to smoke until the oil ceased dripping into the pit sunk in the floor beneath the press. In this interval the next series of cakes went to another press and the work thus kept up during the day.

Six hundred and forty cakes was the average daily output of this family of eight men and two boys, with their six water buffalo.

The cotton seed cakes were being sold as feed, and a near-by Chinese dairyman was using them for his herd of forty water buffalo, seen in Fig. 78, producing milk

Fig. 78.—A dairy herd of water buffalo owned by a Chinese farmer who was supplying milk to foreigners in Shanghai.

for the foreign trade in Shanghai. This herd of forty cows, one of which was an albino, was giving an average of but 200 catty of milk per day, or at the rate of six and two-thirds pounds per head! The cows have extremely small udders but the milk is very rich, as indicated by an analysis made in the office of the Shanghai Board of Health and obtained through the kindness of Dr. Arthur Stanley. The milk showed a specific gravity of 1.028 and contained 20.1 per cent total solids; 7.5 per cent fat; 4.2 per cent milk sugar and .8 per cent ash. In the family of Rev. W. H. Hudson, of the Southern Presby-

terian Mission, Kashing, whose very gracious hospitality
we enjoyed on two different occasions, the butter made
from the milk of two of these cows, one of which, with
her calf, is seen in Fig. 79, was used on the family table.
It was as white as lard or cottolene but the texture and
flavor were normal and far better than the Danish and
New Zealand products served at the hotels.

The milk produced at the Chinese dairy in Shanghai
was being sold in bottles holding two pounds, at the rate
of one dollar a bottle, or 43 cents, gold. This seems high

Fig. 79.—Water buffalo and calf, Kashing, Chekiang province, China.

and there may have been misunderstanding on the part
of my interpreter but his answer to my question was that
the milk was being sold at one Shanghai dollar per bottle
holding one and a half catty, which, interpreted, is the
value given above.

But fuel from the stems of cultivated plants which are
in part otherwise useful, is not sufficient to meet the
needs of country and village, notwithstanding the intense
economies practiced. Large areas of hill and mountain
land are made to contribute their share, as we have seen
in the south of China, where pine boughs were being used

for firing the lime and cement kilns. At Tsingtao we saw the pine bough fuel on the backs of mules, Fig. 80, coming from the hills in Shantung province. Similar fuels were being used in Korea and we have photographs of large pine bough fuel stacks, taken in Japan at Funabashi, east from Tokyo.

Fig. 80.—Pine bough fuel coming into Tsingtao from the Shantung hills, China.

The hill and mountain lands, wherever accessible to the densely peopled plains, have long been cut over and as regularly has afforestation been encouraged and deliberately secured even through the transplanting of nursery stock grown expressly for that purpose. We had read so much regarding the reckless destruction of forests in China and Japan and had seen so few old forest trees except where these had been protected about temples, graves or houses, that when Rev. R. A. Haden, of the Elizabeth Blake hospital, near Soochow, insisted that the Chinese were deliberate foresters and that they regularly grow trees for fuel, transplanting them when necessary to secure a close and early stand, after the area had been cleared, we were so much surprised that he generously volunteered to accompany us westward on a two days journey into the hill country where the practice could be seen.

A family owning a houseboat and living upon it was engaged for the journey. This family consisted of a recently widowed father, his two sons, newly married, and a helper. They were to transport us and provide sleeping quarters for myself, Mr. Haden and a cook for the consideration of $3.00, Mexican, per day and to continue the journey through the night, leaving the day for observation in the hills.

The recent funeral had cost the father $100 and the wedding of the two sons $50 each, while the remodelling of the houseboat to meet the needs of the new family relations cost still another $100. To meet these expenses it had been necessary to borrow the full amount, $300. On $100 the father was paying 20 per cent interest; on $50 he was compelled to pay 50 per cent interest. The balance he had borrowed from friends without interest but with the understanding that he would return the favor should occasion be required.

Rev. A. E. Evans informed us that it is a common practice in China for neighbors to help one another in times of great financial stress. This is one of the methods: A neighbor may need 8000 cash. He prepares a feast and sends invitations to a hundred friends. They know there has been no death in his family and that there is no wedding, still it is understood that he is in need of money. The feast is prepared at a small expense, the invited guests come, each bringing eighty cash as a present. The recipient is expected to keep a careful record of contributing friends and to repay the sum. Another method is like this: For some reason a man needs to borrow 20,000 cash. He proposes to twenty of his friends that they organize a club to raise this sum. If the friends agree each pays 1000 cash to the organizing member. The balance of the club draw lots as to which member shall be number two, three, four, five, etc., designating the order in which payments shall be made. The man borrowing the money is then under obligation to see that these payments are met in full at the times agreed upon. Not infrequently a small rate of interest is charged.

Fig. 81.—Residence houseboat used by family for carrying passengers on rivers and canals, China.

Rates of interest are very high in China, especially on small sums where securities are not the best. Mr. Evans informs me that two per cent per month is low and thirty per cent per annum is very commonly collected. Such obligations are often never met but they do not outlaw and may descend from father to son.

The boat cost $292.40 in U. S. currency; the yearly earning was $107.50 to $120.40. The funeral cost $43 and

$43 more was required for the wedding of the two sons. They were receiving for the services of six people $1.29 per day. An engagement for two weeks or a month could have been made for materially lower rates and their average daily earning, on the basis of three hundred days service in the year, and the $120.40 total earning, would be only 40.13 cents, less than seven cents each, hence their

Fig. 82.—Forest cutting in narrow strips on steep hillsides west of Soochow, China.

trip with us was two of their banner days. Foreigners in Shanghai and other cities frequently engage such houseboat service for two weeks or a month of travel on the canals and rivers, finding it a very enjoyable as well as inexpensive way of having a picnic outing.

On reaching the hill lands the next morning there were such scenes as shown in Fig. 82, where the strips of tree growth, varying from two to ten years, stretched directly up the slope, often in strong contrast on account of the straight boundaries and different ages of the timber. Some of these long narrow holdings were less than two rods wide and on one of these only recently cut, up which we walked for considerable distance, the young pine were springing up in goodly numbers. As many as eighteen young trees were counted on a width of six feet across the strip of thirty feet wide. On this area everything had been recently cut clean. Even stumps and the large roots were dug and saved for fuel.

Fig. 83.—Bundles of pine and oak bough fuel gathered on the hill lands west of Soochow, Kiangsu, China.

In Fig. 83 are seen bundles of fuel from such a strip, just brought into the village, the boughs retaining the leaves although the fuel had been dried. The roots, too, are tied in with the limbs so that everything is saved. On our walk to the hills we passed many people bringing their loads of fuel swinging from carrying poles on their shoulders.

Inquiries regarding the afforestation of these strips of hillside showed that the extensive digging necessitated by the recovery of the roots usually caused new trees to spring up quickly as volunteers from scattered seed and from the roots, so that planting was not generally required. Talking with a group of people as to where we could see some of the trees used for replanting the hillsides, a lad of seven years was first to understand and volunteered to conduct us to a planting. This he did and was overjoyed

Fig. 84.—Tiny nursery of small pines growing among ferns in a shady wood, for replanting cut-over hillsides.

on receipt of a trifle for his services. One of these little pine nurseries is seen in Fig. 84, many being planted in suitable places through the woods. The lad led us to two such locations with whose whereabouts he was evidently very familiar, although they were considerable distance from the path and far from home. These small trees are used in filling in places where the volunteer growth has not been sufficiently close. A strong herbaceous growth usually springs up quickly on these newly cleared lands and this too is cut for fuel or for use in making compost or as green manure.

The grass which grows on the grave lands, if not fed off, is also cut and saved for fuel. We saw several instances of this outside of Shanghai, one where a mother with her daughter, provided with rake, sickle, basket and bag, were gathering the dry stubble and grass of the previous season, from the grave lands where there was less than could

Fig. 85.—Dried grass fuel gathered on grave lands, Shanghai.

be found on our closely mowed meadows. In Fig. 85 may be seen a man who has just returned with such a load, and in his hand is the typical rake of the Far East, made by simply bending bamboo splints, claw-shape, and securing them as seen in the engraving.

In the Shantung province, in Chihli and in Manchuria, millet stems, especially those of the great kaoliang or sorghum, are extensively used for fuel and for building as well as for screens, fences and matting. At Mukden the kaoliang was selling as fuel at \$2.70 to \$3.00, Mexican, for a 100-bundle load of stalks, weighing seven catty to the bundle. The yield per acre of kaoliang fuel amounts to 5600 pounds and the stalks are eight to twelve feet long, so that when carried on the backs of mules or horses the animals are nearly hidden by the load. The price paid

Fig. 86.—Bundles of kaoliang fuel coming into Kiaochow market, Shuntung.

for plant stem fuel from agricultural crops, in different parts of China and Japan, ranged from \$1.30 to \$2.85, U. S. currency, per ton. The price of anthracite coal at Nanking was \$7.76 per ton. Taking the weight of dry oak wood at 3500 pounds per cord, the plant stem fuel, for equal weight, was selling at \$2.28 to \$5.00.

Large amounts of wood are converted into charcoal in these countries and sent to market baled in rough matting or in basketwork cases woven from small brush and holding two to two and a half bushels. When such wood is not converted into charcoal it is sawed into one or two-foot lengths, split and marketed tied in bundles, as seen in Fig. 77.

Along the Mukden–Antung railway in Manchuria fuel was also being shipped in four-foot lengths, in the form of cordwood. In Korea cattle were provided with a peculiar saddle for carrying wood in four-foot sticks laid blanket-fashion over the animal, extending far down on their sides. Thus was it brought from the hills to the railway station. This wood, as in Manchuria, was cut from small trees. In Korea, as in most parts of China where we visited, the tree growth over the hills was generally scattering and thin on the ground wherever there was not individual ownership in small holdings. Under and among the scattering pine there were oak in many cases, but these were always small, evidently not more than two or three years standing, and appearing to have been repeatedly cut back. It was in Korea that we saw so many instances of young leafy oak boughs brought to the rice fields and used as green manure.

There was abundant evidence of periodic cutting between Mukden and Antung in Manchuria; between Wiju and Fusan in Korea; and throughout most of our journey in Japan; from Nagasaki to Moji and from Shimonoseki to Yokohama. In all of these countries afforestation takes place quickly and the cuttings on private holdings are made once in ten, twenty or twenty-five years. When the wood is sold to those coming for it the takers pay at the rate of 40 sen per one horse load of forty kan, or 330 pounds, such as is seen in Fig. 87. Director Ono, of the Akashi Experiment station, informed us that such fuel loads in that prefecture, where the wood is cut once in ten years, bring returns amounting to about $40 per acre for the ten-year crop. This land was worth $40 per acre but when they are suitable for orange groves they sell for $600 per acre. Mushroom culture is extensively practiced under the shade of some of these wooded areas, yielding under favorable conditions at the rate of $100 per acre.

The forest covered area in Japan exclusive of Formosa and Karafuto, amounts to a total of 54,196,728 acres, less than twenty millions of which are in private holdings, the balance belonging to the state and to the Imperial Crown.

In all of these countries there has been an extensive

Fig. 87.—Japanese fuel coming down from the wooded hills.

general use of materials other than wood for building purposes and very many of the substitutes for lumber are products grown on the cultivated fields. The use of rice straw for roofing, as seen in the Hakone village, Fig. 8, is very general throughout the rice growing districts, and even the sides of houses may be similarly thatched, as was observed in the Canton delta region, such a construction being warm for winter and cool for summer. The life of these thatched roofs, however, is short and they must be renewed as often as every three to five years but the old straw is highly prized as fertilizer for the fields on which it is grown, or it may serve as fuel, the ashes only going to the fields.

Burned clay tile, especially for the cities and public buildings, are very extensively used for roofing, clay being abundant and near at hand. In Chihli and in Manchuria millet and sorghum stems, used alone or plastered, as in Fig. 88, with a mud mortar, sometimes mixed with

Fig. 88.—Millet-thatched roofs plastered with earth; mud chimneys; walls of houses plastered with mud, and winter storage pits for vegetables built of clay and chaff mortar.

lime, cover the roofs of vast numbers of the dwellings outside the larger cities.

At Chiao Tou in Manchuria we saw the building of the thatched millet roofs and the use of kaoliang stems as lumber. Rafters were set in the usual way and covered with a layer about two inches thick of the long kaoliang stems stripped of their leaves and tops. These were tied together and to the rafters with twine, thus forming a sort of matting. A layer of thin clay mortar was then spread over the surface and well trowelled until it began to show on the under side. Over this was applied a thatch of small millet stems bound in bundles eight inches thick, cut

11

square across the butts to eighteen inches in length. They were dipped in water and laid in courses after the manner of shingles but the butts of the stems are driven forward to a slope which obliterates the shoulder, making the courses invisible. In the better houses this thatching may be plastered with earth mortar or with an earth-lime mortar, which is less liable to wash in heavy rain.

Fig. 89.—Air-dried earth brick for house building.

The walls of the house we saw building were also sided with the long, large kaoliang stems. An ordinary frame with posts and girts about three feet apart had been erected, on sills and with plates carrying the roof. Standing vertically against the girts and tied to them, forming a close layer, were the kaoliang stems. These were plastered outside and in with a layer of thin earth mortar. A similar layer of stems, set up on the inside of the girts and similarly plastered, formed the inner face of the wall of the house, leaving dead air spaces between the girts.

Brick made from earth are very extensively used for house building, chaff and short straw being used as a binding material, the brick being simply dried in the sun, as seen in Fig. 89. A house in the process of building, where the brick were being used, is seen in Fig. 90. The foundation of the dwelling, it will be observed, was laid with well-formed hard-burned brick, these being necessary to prevent capillary moisture from the ground being drawn up and soften the earth brick, making the wall unsafe.

Fig. 90.—Foundation of dwelling, consisting of hard-burned brick; balance of wall to be sun-dried earth brick, seen in Fig. 89.

Several kilns for burning brick, built of clay and earth, were passed in our journey up the Pei ho, and stacked about them, covering an area of more than eight hundred feet back from the river, were bundles of the kaoliang stems to serve as fuel in the kilns.

The extensive use of the unburned brick is necessitated by the difficulty of obtaining fuel, and various methods are adopted to reduce the number of burned brick required in construction. One of these devices is shown in Fig. 79, where the city wall surrounding Kashing is constructed of alternate courses of four layers of burned brick separated by layers of simple earth concrete.

In addition to the multiple-function, farm-grown crops used for food, fuel and building material, there is a large acreage devoted to the growing of textile and fiber products and enormous quantities of these are produced annually. In Japan, where some fifty millions of people are chiefly fed on the produce of little more than 21,000 square miles of cultivated land, there was grown in 1906 more than 75,500,000 pounds of cotton, hemp, flax and China grass

Fig. 91.—Earth and clay brick kiln on the bank of the Pei ho, using sorghum stems for fuel.

textile stock, occupying 76,700 acres of the cultivated land. On 141,000 other acres there grew 115,000,000 pounds of paper mulberry and Mitsumata, materials used in the manufacture of paper. From still another 14,000 acres were taken 92,000,000 pounds of matting stuff, while more than 957,000 acres were occupied by mulberry trees for the feeding of silkworms, yielding to Japan 22,389,798 pounds of silk. Here are more than 300,000,000 pounds of fiber and textile stuff taken from 1860 square miles of the cultivated land, cutting down the food producing area to

19,263 square miles and this area is made still smaller by devoting 123,000 acres to tea, these producing in 1906 58,900,000 pounds, worth nearly five million dollars. Nor do these statements express the full measure of the producing power of the 21,321 square miles of cultivated land, for, in addition to the food and other materials named, there were also made $2,365,000 worth of braid from straw and wood shavings; $6,000,000 worth of rice straw bags, packing cases and matting; and $1,085,000 worth of wares from bamboo, willow and vine. As illustrating the intense home industry of these people we may consider the fact that the 5,453,309 households of farmers in Japan produced in 1906, in their homes as subsidiary work, $20,527,000 worth of manufactured articles. If correspondingly exact statistical data were available from China and Korea a similarly full utilization of cultural possibilities would be revealed there.

This marvelous heritage of economy, industry and thrift, bred of the stress of centuries, must not be permitted to lose virility through contact with western wasteful practices, now exalted to seeming virtues through the dazzling brilliancy of mechanical achievements. More and more must labor be dignified in all homes alike, and economy, industry and thrift become inherited impulses compelling and satisfying.

Cheap, rapid, long distance transportation, already well started in these countries, will bring with it a fuller utilization of the large stores of coal and mineral wealth and of the enormous available water power, and as a result there will come some temporary lessening of the stress for fuel and with better forest management some relief along the lines of building materials. But the time is not a century distant when, throughout the world, a fuller, better development must take place along the lines of these most far-reaching and fundamental practices so long and so effectively followed by the Mongolian races in China, Korea and Japan. When the enormous water-power of these countries has been harnessed and brought into the foot-hills and down upon the margins of the valleys and plains in the form of electric current, let it, if possible, be in a large measure so distributed as to become available in the country village homes to lighten the burden and lessen the human drudgery and yet increase the efficiency of the human effort now so well bestowed upon subsidiary manufactures under the guidance and initiative of the home, where there may be room to breathe and for children to come up to manhood and womanhood in the best conditions possible, rather than in enormous congested factories.

VIII.

TRAMPS AFIELD.

On March 31st we took the 8 A. M. train on the Shang-hai-Nanking railway for Kunshan, situated thirty-two miles west from Shanghai, to spend the day walking in the fields. The fare, second class, was eighty cents, Mexican. A third class ticket would have been forty cents and a first class, $1.60, practically two cents, one cent and half a cent, our currency, per mile. The second class fare to Nanking, a distance of 193 miles, was $1.72, U. S. currency, or a little less than one cent per mile. While the car seats were not upholstered, the service was good. Meals were served on the train in either foreign or Chinese style, and tea, coffee or hot water to drink. Hot, wet face cloths were regularly passed and many Chinese daily newspapers were sold on the train, a traveler often buying two.

In the vicinity of Kunshan a large area of farm land had been acquired by the French catholic mission at a purchase price of $40, Mexican, per mow, or at the rate of $103.20 per acre. This they rented to the Chinese.

It was here that we first saw, at close range, the details of using canal mud as a fertilizer, so extensively applied in China. Walking through the fields we came upon the scene in the middle section of Fig. 92 where, close on the right was such a reservoir as seen in Fig. 58. Men were in it, dipping up the mud which had accumulated over its bottom, pouring it on the bank in a field of windsor beans, and the thin mud was then over two feet deep at that side

Fig. 92.—In the lower section, along the path, basketsful of canal mud had been applied in two rows at the rate of more than 100 tons per acre. In the middle section workmen just beyond the extreme right were removing mud from such a reservoir as is seen in Fig. 58. The upper section shows three men distributing canal mud between the rows of a field of windsor beans.

and flowing into the beans where it had already spread two rods, burying the plants as the engraving shows. When sufficiently dry to be readily handled this would be spread among the beans as we found it being done in another field, shown in the upper section of the illustration. Here four men were distributing such mud, which had dried, between the rows, not to fertilize the beans, but for a succeeding crop of cotton soon to be planted between the rows, before they were harvested. The owner of this piece of land, with whom we talked and who was superintending the work, stated that his usual yield of these beans was three hundred catty per mow and that they sold them green, shelled, at two cents, Mexican, per catty. At this price and yield his return would be $15.48, gold, per acre. If there was need of nitrogen and organic matter in the soil the vines would be pulled green, after picking the beans, and composted with the wet mud. If not so needed the dried stems would be tied in bundles and sold as fuel or used at home, the ashes being returned to the fields. The windsor beans are thus an early crop grown for fertilizer, fuel and food.

This farmer was paying his laborers one hundred cash per day and providing their meals, which he estimated worth two hundred cash more, making twelve cents, gold, for a ten-hour day. Judging from what we saw and from the amount of mud carried per load, we estimated the men would distribute not less than eighty-four loads of eighty pounds each per day, an average distance of five hundred feet, making the cost 3.57 cents, gold, per ton for distribution.

The lower section of Fig. 92 shows another instance where mud was being used on a narrow strip bordering the path along which we walked, the amount there seen having been brought more than four hundred feet, by one man before 10 A. M. on the morning the photograph was taken. He was getting it from the bottom of a canal ten feet deep, laid bare by the out-going tide. Already he had brought more than a ton to his field.

The carrying baskets used for this work were in the form of huge dustpans suspended from the carrying poles by two cords attached to the side rims, and steadied by the hand grasping a handle provided in the back for this purpose and for emptying the baskets by tipping. With this construction the earth was readily raked upon the basket and very easily emptied from it by simply raising the hands when the destination was reached. No arrangement could be more simple, expeditious or inexpensive for this man with his small holding. In this simple manner has nearly all of the earth been moved in digging the miles of canal and in building the long sea walls. In Shanghai the mud carried through the storm sewers into Soochow creek we saw being removed in the same manner during the intervals when the tide was out.

In still another field, seen in Fig. 93, the upper portion shows where canal mud had been applied at a rate exceeding seventy tons per acre, and we were told that such dressings may be repeated as often as every two years though usually at longer intervals, if other and cheaper fertilizers could be obtained. In the lower portion of the same illustration may be seen the section of canal from which this mud was taken up the three earthen stairways built of the mud itself and permitted to dry before using. Many such lines of stairway were seen during our trips along the canals, only recently made or in the process of building to be in readiness when the time for applying the mud should arrive. To facilitate collecting the mud from the shallow canals temporary dams may be thrown across them at two places and the water between either scooped or pumped out, laying the bottom bare, as is often done also for fishing. The earth of the large grave mound seen across a canal in the center background of the upper portion of the engraving had been collected in a similar manner.

In the Chekiang province canal mud is extensively used in the mulberry orchards as a surface dressing. We have referred to this practice in southern China, and Fig. 94 is a view taken south of Kashing early in April. The boat

anchored in front of the mulberry orchard is the home of
a family coming from a distance, seeking employment during the season for picking mulberry leaves to feed silkworms. We were much surprised, on looking back at the

Fig. 93.—Section of field covered with piles of canal mud recently applied at
the rate of more than 70 tons per acre; taken out of the canal up the three
flights of earth steps shown in the lower part of the figure.

boat after closing the camera, to see the head of the family
standing erect in the center, having shoved back a section
of the matting roof.

The dressing of mud applied to this field formed a loose
layer more than two inches deep and when compacted by
the rains which would follow would add not less than a full

Fig. 94.—Mulberry orchard to which a heavy dressing of canal mud had been applied. A family of mulberry leaf pickers were living in the boat anchored in the canal.

inch of soil over the entire orchard, and the weight per acre could not be less than 120 tons.

Another equally, or even more, laborious practice followed by the Chinese farmers in this province is the periodic exchange of soil between mulberry orchards and the rice fields, their experience being that soil long used in the mulberry orchards improves the rice, while soil from the rice fields is very helpful when applied to the mulberry orchards. We saw many instances, when travelling by boat-train between Shanghai, Kashing and Hangchow, of soil being carried from rice fields and either stacked on the banks or dropped into the canal. Such soil was oftenest taken from narrow trenches leading through the fields, laying them off in beds. It is our judgment that the soil thrown into the canals undergoes important changes, perhaps through the absorption of soluble plant food substances such as lime, phosphoric acid and potash withdrawn from the water, or through some growth or fermentation, which, in the judgment of the farmer, makes the large labor involved in this procedure worth while. The stacking of soil along the banks was probably in preparation for its removal by boat to some of the mulberry orchards.

It is clearly recognized by the farmers that mud collected from those sections of the canal leading through country villages, such as that seen in Fig. 10, is both inherently more fertile and in better physical condition than that collected in the open country. They attribute this difference to the effect of the village washing in the canal, where soap is extensively used. The storm waters of the city doubtless carry some fertilizing material also, although sewage, as such, never finds its way into the canals. The washing would be very likely to have a decided flocculating effect and so render this material more friable when applied to the field.

One very important advantage which comes to the fields when heavily dressed with such mud is that resulting from the addition of lime which has become incorporated with the silts through their flocculation and precipitation, and that which is added in the form of snail shells abounding in the canals. The amount of these may be realized from the

Fig. 95.—The recently removed canal mud, in the upper section of the illustration, is heavily charged with large snail shells. The lower section shows the shells in the soil of a recently spaded field.

large numbers contained in the mud recently thrown out, as seen in the upper section of Fig. 95, where the pebbly appearance of the surface is caused by snail shells. In the lower section of the same illustration the white spots are snail shells exposed in the soil of a recently spaded field. The shells are by no means as numerous generally as here seen but yet sufficient to maintain the supply of lime.

Several species of these snails are collected in quantities and used as food. Piles containing bushels of the empty shells were seen along the canals outside the villages. The snails are cooked in the shell and often sold by measure to be eaten from the hand, as we buy roasted peanuts or popcorn. When a purchase is made the vender clips the spiral point from each shell with a pair of small shears. This admits air and permits the snail to be readily removed by suction when the lips are applied to the shell. In the canals there are also large numbers of fresh water eel, shrimp and crabs as well as fish, all of which are collected and used for human food. It is common, when walking through the canal country, to come upon groups of gleaners busy in the bottoms of the shallow agricultural canals, gathering anything which may serve as food, even including small bulbs or the fleshy roots of edible aquatic plants. To facilitate the collection of such food materials sections of the canal are often drained in the manner already described, so that gleaning may be done by hand, wading in the mud. Families living in houseboats make a business of fishing for shrimp. They trail behind the houseboat one or two other boats carrying hundreds of shrimp traps cleverly constructed in such manner that when they are trailed along the bottom and disturb the shrimps they dart into the holes in the trap, mistaking them for safe hiding places.

On the streets, especially during festival days, one may see young people and others in social intercourse, busying their fingers and their teeth eating cooked snails or often watermelon seeds, which are extensively sold and thus eaten. This custom we saw first in the streets of a city south of Kashing on the line of the new railway between Hangchow and Shanghai. The first passenger train over the line had been run the day before our visit, which was a festival day and throngs of people were visiting the nine-story pagoda standing on a high hill a mile outside the city limits. The day was one of great surprises to these people who had never before seen a passenger train, and my own person appeared to be a great curiosity to many. No boy ever scrutinized the face of a caged chimpanzee closer, with purer curiosity, or with less consideration for his feelings than did a woman of fifty scrutinize mine, standing close in front, not two feet distant, even bending forward as I sat upon a bench writing at the railway station. People would pass their hands along my coat sleeve to judge the cloth, and a boy felt of my shoes. Walking through the street we passed many groups gathered about tables and upon seats, visiting or in business conference, their fingers occupied with watermelon seeds or with packages of cooked snails. Along the pathway leading to the pagoda beggars had distributed themselves, one in a place, at intervals of two or three hundred feet, asking alms, most of them infirm with age or in some other way physically disabled. We saw but one who appeared capable of earning a living.

Travel between Shanghai and Hangchow at this time was heavy. Three companies were running trains, of six or more houseboats, each towed by a steam launch, and these were daily crowded with passengers. Our train left Shanghai at 4:30 P. M., reaching Hangchow at 5:30 P. M. the following day, covering a distance along the canal of something more than 117 miles. We paid $5.16, gold, for the exclusive use of a first-cabin, five-berth stateroom for myself and intepreter. It occupied the full width of the boat, lacking about fourteen inches of footway, and could be entered from either side down a flight of five steps. The berths were flat, naked wooden shelves thirty inches wide, separated by a partition headboard six inches high and without railing in front. Each traveler provided his own bedding. A small table upon which meals were served, a mirror on one side and a lamp on the other, set in an opening in the partition, permitting it to serve two staterooms, completed the furnishings. The roof of the staterooms was covered with an awning and divided crosswise into two tiers of berths, each thirty inches wide, by board partitions six inches high. In these sections passengers spread their beds, sleeping heads together, separated only by a headboard six inches high. The awning was only sufficiently high to permit passengers to sit erect. Ventilation was ample but privacy was nil. Curtains could be dropped around the sides in stormy weather.

Meals were served to each passenger wherever he might be. Dinner consisted of hot steamed rice brought in very heavy porcelain bowls set inside a covered, wet, steaming hot wooden case. With the rice were tiny dishes, butter-chip size, of green clover, nicely cooked and seasoned; of cooked bean curd served with shredded bamboo sprouts; of tiny pork strips with bean curd; of small bits of liver with bamboo sprouts; of greens, and hot water for tea. If the appetite is good one may have a second helping of rice and as much hot water for tea as desired. There was no table linen, no napkins and everything but the tea had to be negotiated with chop sticks, or, these failing, with the fingers. When the meal was finished the table was cleared and water, hot if desired, was brought for your hand basin, which with tea, teacup and bedding, constitute part of the traveler's outfit. At frequent intervals, up to ten P. M., a crier walked about the deck with hot water for those who might desire an extra cup of tea, and again in the early morning.

At this season of the year Chinese incubators were being run to their full capacity and it was our good fortune to visit one of these, escorted by Rev. R. A. Haden, who also acted as interpreter. The art of incubation is very old and very extensively practiced in China. An interior view of one of these establishments is shown in Fig. 96, where the family were hatching the eggs of hens, ducks and geese, purchasing the eggs and selling the young as hatched. As in the case of so many trades in China, this family was the last generation of a long line whose lives had been spent in the same work. We entered through their store, opening on the street of the narrow village seen in Fig. 10. In the store the eggs were purchased and the chicks were

12

sold, this work being in charge of the women of the family. It was in the extreme rear of the home that thirty incubators were installed, all doing duty and each having a capacity of 1,200 hens' eggs. Four of these may be seen in the illustration and one of the baskets which, when two-thirds filled with eggs, is set inside of each incubator.

Each incubator consists of a large earthenware jar having a door cut in one side through which live charcoal may

Fig. 96.—Four Chinese incubators in a room where there are thirty, each having a capacity of 1,200 hens' eggs.

be introduced and the fire partly smothered under a layer of ashes, this serving as the source of heat. The jar is thoroughly insulated, cased in basketwork and provided with a cover, as seen in the illustration. Inside the outer jar rests a second of nearly the same size, as one teacup may in another. Into this is lowered the large basket with its 600 hens' eggs, 400 ducks' eggs or 175 geese' eggs, as the case may be. Thirty of these incubators were arranged in two parallel rows of fifteen each. Immediately above each row, and utilizing the warmth of the air rising from them, was a continuous line of finishing hatchers and brooders in the form of woven shallow trays with sides warmly padded with cotton and with the tops covered with sets of quilts of different thickness.

After a basket of hens' eggs has been incubated four days it is removed and the eggs examined by lighting, to remove those which are infertile before they have been rendered unsalable. The infertile eggs go to the store and the basket is returned to the incubator. Ducks' eggs are similarly examined after two days and again after five days incubation; and geese' eggs after six days and again after fourteen days. Through these precautions practically all loss from infertile eggs is avoided and from 95 to 98 per cent of the fertile eggs are hatched, the infertile eggs ranging from 5 to 25 per cent.

After the fourth day in the incubator all eggs are turned five times in twenty-four hours. Hens' eggs are kept in the lower incubator eleven days; ducks' eggs thirteen days, and geese' eggs sixteen days, after which they are transferred to the trays. Throughout the incubation period the most careful watch and control is kept over the temperature. No thermometer is used but the operator raises the lid or quilt, removes an egg, pressing the large end into the eye socket. In this way a large contact is made where the skin is sensitive, nearly constant in temperature, but little below blood heat and from which the air is excluded for the time. Long practice permits them thus to judge small differences of temperature expeditiously and with great accuracy; and they maintain different temperatures during different stages of the incubation. The men sleep in the room and some one is on duty continuously, making the rounds of the incubators and brooders, examining and regulating each according to its individual needs, through the management of the doors or the shifting of the quilts over the eggs in the brooder trays where the chicks leave the eggs and remain until they go to the store. In the finishing trays the eggs form rather more than one continuous layer but the second layer does not cover more than a fifth or a quarter of the area. Hens' eggs are in these trays ten days, ducks' and geese' eggs, fourteen days.

After the chickens have been hatched sufficiently long to require feeding they are ready for market and are then sorted according to sex and placed in separate shallow woven trays thirty inches in diameter. The sorting is done rapidly and accurately through the sense of touch, the operator recognizing the sex by gently pinching the anus. Four trays of young chickens were in the store fronting on the street as we entered and several women were making purchases, taking five to a dozen each. **Dr.** Haden informed me that nearly every family in the cities, and in the country villages raise a few, but only a few, chickens and it is a common sight to see grown chickens walking about the narrow streets, in and out of the open stores, dodging the feet of the occupants and passers by. At the time of our visit this family was paying at the rate of ten cents, Mexican, for nine hens' and eight ducks' eggs, and were selling their largest strong chickens at three cents each. These figures, translated into our currency, make the purchase price for eggs nearly 48 cents, and the selling price for the young chicks $1.29, per hundred, or thirteen eggs for six cents and seven chickens for nine cents.

It is difficult even to conceive, not to say measure, the vast import of this solution of how to maintain, in the millions of homes, a constantly accessible supply of absolutely fresh and thoroughly sanitary animal food in the form of meat and eggs. The great density of population in these countries makes the problem of supplying eggs to the people very different from that in the United States. Our 250,600,000 fowl in 1900 was at the rate of three to each person but in Japan, with her 16,500,000 fowl, she had in 1906 but one for every three people. Her number per square mile of cultivated land however was 825, while in the United States, in 1900, the number of fowls per square mile of improved farm land was but 387. To give to Japan three fowls to each person there would needs be an average of about nine to each acre of her cultivated land, whereas in the United States there were in 1900 nearly two acres of improved farm land for each fowl. We have no statistics regarding the number of fowl in China or the number of eggs produced but the total is very large and she exports to Japan. The large boat load of eggs seen in Fig. 97 had just arrived from the country, coming into Shanghai in one of her canals.

Besides applying canal mud directly to the fields in the ways described there are other very extensive practices of composting it with organic matter of one or another kind

Fig. 97.—Boatload of 150 baskets of eggs on Soochow creek, Shanghai, China.

and of then using the compost on the fields. The next three illustrations show some of the steps and something of the tremendous labor of body, willingly and cheerfully incurred, and something of the forethought practiced, that homes may be maintained and that grandparents, parents, wives and children need neither starve nor beg. We had reached a place seen in Fig. 98, where eight bearers were moving winter compost to a recently excavated pit in an adjoining field shown in Fig. 99.

Four months before the camera fixed the activity shown, men had brought waste from the stables of Shanghai fifteen miles by water, depositing it upon the canal bank between layers of thin mud dipped from the canal, and left it to ferment. The eight men were removing this compost to the pit seen in Fig. 99, then nearly filled. Near by in the same field was a second pit seen in Fig. 100, excavated three feet deep and rimmed about with the earth removed, making it two feet deeper.

Fig. 98.—Eight bearers moving a pile of winter compost to the recently excavated pit in the field seen in Fig. 99. The boatload in the foreground is a mixture of manure and ashes just arrived from the home village.

After these pits had been filled the clover which was in blossom beyond the pits would be cut and stacked upon them to a hight of five to eight feet and this also saturated, layer by layer, with mud brought from the canal, and allowed to ferment twenty to thirty days until the juices set free had been absorbed by the winter compost beneath, helping to carry the ripening of that still further, and until the time had arrived for fitting the ground for the next crop. This organic matter, fermented with the canal mud, would then be distributed by the men over the field, carried a third time on their shoulders, notwithstanding its weight was many tons.

Fig. 99.—Compost pit adjacent to a field of clover, being filled from the pile of winter compost seen on the bank of the canal in Fig. 98.

Fig. 100.—Recently excavated pit for receiving winter compost seen in Fig. 98, and upon which the clover beyond the pit will be cut and composted as a fertilizer for a crop of rice.

This manure had been collected, loaded and carried fifteen miles by water; it had been unloaded upon the bank and saturated with canal mud; the field had been fitted for clover the previous fall and seeded; the pits had been dug in the fields; the winter compost had been carried and placed in the pits; the clover was to be cut, carried by the men on their shoulders, stacked layer by layer and saturated with mud dipped from the canal; the whole would later be distributed over the field and finally the earth removed from the pits would be returned to them, that the service

Fig. 101.—Providing for the building of a mud-and-clover compost stack.

of no ground upon which a crop might grow should be lost.

Such are the tasks to which Chinese farmers hold themselves, because they are convinced desired results will follow, because their holdings are so small and their families so large. These practices are so extensive in China and so fundamental in the part they play in the maintenance of high productive power in their soils that we made special effort to follow them through different phases. In Fig. 101 we saw the preparation being made to build one of the clover compost stacks saturated with canal mud. On the left the thin mud had been dipped from the canal; way-farers in the center were crossing the foot-bridge of the

Fig. 102.—Clover compost stack in the building.

country by-way; and beyond rises the conical thatch to
shelter the water buffalo when pumping for irrigating the
rice crop to be fed with this plant food in preparation.   On
the right were two large piles of green clover freshly cut
and a woman of the family at one of them was spreading
it to receive the mud, while the men-folk were coming from
the field with more clover on their carrying poles.   We
came upon this scene just before the dinner hour and after

Fig. 103.—The young man is loading his boat with canal mud, using the long-
handled clam-shell dredge which he can open and close at will.

the workers had left another photograph was taken at
closer range and from a different side, giving the view seen
in Fig. 102.   The mud had been removed some days and
become too stiff to spread, so water was being brought from
the canal in the pails at the right for reducing its consistency to that of a thin porridge, permitting it to more
completely smear and saturate the clover.   The stack
grew, layer by layer, each saturated with the mud, tramped
solid with the bare feet, trousers rolled high.   Provision
had been made here for building four other stacks.

Further along we came upon the scene in Fig. 103 where

the building of the stack of compost and the gathering of the mud from the canal were simultaneous. On one side of the canal the son, using a clam-shell form of dipper made of basket-work, which could be opened and shut with a pair of bamboo handles, had nearly filled the middle section of his boat with the thin ooze, while on the other side, against the stack which was building, the mother was emptying a similar boat, using a large dipper, also provided with a bamboo handle. The man on the stack is a good scale for judging its size.

Fig. 104.—A completed compost stack.

We came next upon a finished stack on the bank of another canal, shown in Fig. 104, where our umbrella was set to serve as a scale. This stack measured ten by ten feet on the ground, was six feet high and must have contained more than twenty tons of the green compost. At the same place, two other stacks had been started, each about fourteen by fourteen feet, and foundations were laid for six others, nine in all.

During twenty or more days this green nitrogenous organic matter is permitted to lie fermenting in contact with the fine soil particles of the ooze with which it had been charged. This is a remarkable practice in that it is a very old, intensive application of an important fundamental principle only recently understood and added to the science of agriculture, namely, the power of organic matter, decaying rapidly in contact with soil, to liberate from it soluble plant food; and so it would be a great mistake to say that these laborious practices are the result of ignorance, of a lack of capacity for accurate thinking or of power to grasp and utilize. If the agricultural lands of the United States are ever called upon to feed even 1200 millions of people, a number proportionately less than one-half that being fed in Japan today, very different practices from those we are now following will have been adopted. We can believe they will require less human bodily effort and be more efficient. But the knowledge which can make them so is not yet in the possession of our farmers, much less the conviction that plant feeding and more persistent and better directed soil management are necessary to such yields as will then be required.

Later, just before the time for transplanting rice, we returned to the same district to observe the manner of applying this compost to the field, and Fig. 105 is prepared from photographs taken then, illustrating the activities of one family, as seen during the morning of May 28th. Their home was in a near-by village and their holding was divided into four nearly rectangular paddies, graded to water level, separated by raised rims, and having an area of nearly two acres. Three of these little fields are partly shown in the illustration, and the fourth in Fig. 160. In the background of the upper section of Fig. 105, and under the thatched shelter, was a native Chinese cow, blindfolded and hitched to the power-wheel of a large wooden-chain pump, lifting water from the canal and flooding the field in the foreground, to soften the soil for plowing. Riding on the power-wheel was a girl of some twelve years, another of seven and a baby. They were there for entertainment and to see that the cow kept at work. The ground had been sufficiently softened so that the father had begun plowing, the cow sinking to her knees as she walked. In the same paddy, but shown in the section below, a boy was spreading the clover compost with his hands, taking care that it was finely divided and evenly scattered. He had been once around before the plowing began. This compost had been brought from a stack by the side of a canal, and two other men were busy still bringing the material to one of the other paddies, one of whom, with his baskets on the carrying pole appears in the third section. Between these two paddies was the one seen at the bottom of the illustration, which had matured a crop of rape that had been pulled and was lying in swaths ready to be moved. Two other men were busy here, gathering the rape into large bundles and carrying it to the village home, where the women were threshing out the seed, taking care not to break the stems which, after threshing, were tied into bundles for fuel. The seed would be ground and from it an oil expressed, while the cake would be used as a fertilizer.

This crop of rape is remarkable for the way it fits into the economies of these people. It is a near relative of mustard and cabbage; it grows rapidly during the cooler portions of the season, the spring crop ripening before the planting of rice and cotton; its young shoots and leaves are succulent, nutritious, readily digested and extensively used as human food, boiled and eaten fresh, or salted for winter use, to be served with rice; the mature stems, being woody, make good fuel; and it bears a heavy crop of seed, rich in oil, which has been extensively used for lights and in cooking, while the rape seed cake is highly prized as a manure and very extensively so used.

In the early spring the country is luxuriantly green with the large acreage of rape, later changing to a sea of most brilliant yellow and finally to an ashy grey when the leaves fall and the stems and pods ripen. Like the dairy cow, rape produces a fat, in the ratio of about forty pounds of oil to a hundred pounds of seed, which may be eaten, burned or sold without materially robbing the soil of its

Fig. 195.—The activities of a family, fertilizing and fitting paddies for rice.

fertility if the cake and the ashes from the stems are returned to the fields, the carbon, hydrogen and oxygen of which the oil is almost wholly composed coming from the atmosphere rather than from the soil.

In Japan rape is grown as a second crop on both the upland and paddy fields, and in 1906 she produced more than 5,547,000 bushels of the seed; $1,845,000 worth of rape seed cake, importing enough more to equal a total value of $2,575,000, all of which was used as a fertilizer, the oil being exported. The yield of seed per acre in Japan ranges between thirteen and sixteen bushels, and the farmer whose field was photographed estimated that his returns from the crop would be at the rate of 640 pounds of seed per acre, worth $6.19, and 8,000 pounds of stems worth as fuel $5.16 per acre.

IX.

THE UTILIZATION OF WASTE.

One of the most remarkable agricultural practices adopted by any civilized people is the centuries-long and well nigh universal conservation and utilization of all human waste in China, Korea and Japan, turning it to marvelous account in the maintenance of soil fertility and in the production of food. To understand this evolution it must be recognized that mineral fertilizers so extensively employed in modern western agriculture, like the extensive use of mineral coal, had been a physical impossibility to all people alike until within very recent years. With this fact must be associated the very long unbroken life of these nations and the vast numbers their farmers have been compelled to feed.

When we reflect upon the depleted fertility of our own older farm lands, comparatively few of which have seen a century's service, and upon the enormous quantity of mineral fertilizers which are being applied annually to them in order to secure paying yields, it becomes evident that the time is here when profound consideration should be given to the practices the Mongolian race has maintained through many centuries, which permit it to be said of China that one-sixth of an acre of good land is ample for the maintenance of one person, and which are feeding an average of three people per acre of farm land in the three southernmost of the four main islands of Japan.

From the analyses of mixed human excreta made by Wolff in Europe and by Kellner in Japan it appears that, as an average, these carry in every 2000 pounds 12.7 pounds of nitrogen, 4 pounds of potassium and 1.7 pounds of phosphorus. On this basis and that of Carpenter, who estimates the average amount of excreta per day for the adult at 40 ounces, the average annual production per million of adult population is 5,794,300 pounds of nitrogen; 1,825,000 pounds of potassium, and 775,600 pounds of phosphorus carried in 456,250 tons of excreta. The figures which Hall cites in Fertilizers and Manures, would make these amounts 7,940,000 pounds of nitrogen; 3,070,500 pounds of potassium, and 1,965,600 pounds of phosphorus. but the figures he takes and calls high averages give 12,000,000 of nitrogen; 4,151,000 pounds of potassium, and 3,057,600 pounds of phosphorus.

In 1908 the International Concessions of the city of Shanghai sold to one Chinese contractor for $31,000, gold, the privilege of collecting 78,000 tons of human waste, under stipulated regulations, and of removing it to the country for sale to farmers. The flotilla of boats seen in Fig. 106 is one of several engaged daily in Shanghai throughout the year in this service.

Dr. Kawaguchi, of the National Department of Agriculture and Commerce, taking his data from their records, informed us that the human manure saved and applied to the fields of Japan in 1908 amounted to 23,850,295 tons, which is an average of 1.75 tons per acre of their 21,321 square miles of cultivated land in their four main islands.

On the basis of the data of Wolff, Kellner and Carpenter, or of Hall, the people of the United States and of Europe are pouring into the sea, lakes or rivers and into the underground waters from 5,794,300 to 12,000,000 pounds of nitrogen; 1,881,900 to 4,151,000 pounds of potassium, and 777,200 to 3,057,600 pounds of phosphorus per million of adult population annually, and this waste we esteem one of the great achievements of our civilization. In the Far East, for more than thirty centuries, these enormous wastes have been religiously saved and today the four hundred million of adult population send back to their fields annually 150,000 tons of phosphorus; 376,000

Fig. 106.—A flotilla of manure boats on Soochow creek, collecting human wastes in the city of Shanghai, China, for removal to cultivated fields.

tons of potassium, and 1,158,000 tons of nitrogen comprised in a gross weight exceeding 182 million tons, gathered from every home, from the country villages and from the great cities like Hankow-Wuchang-Hanyang with its

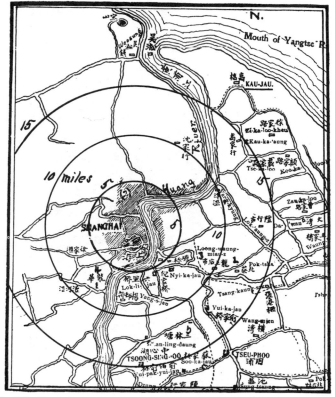

Fig. 107.—Map of country surrounding Shanghai, China, showing a few of the many canals on which the waste of the city is conveyed by boat to the farms.

1,770,000 people swarming on a land area delimited by a radius of four miles.

Man is the most extravagant accelerator of waste the world has ever endured. His withering blight has fallen upon every living thing within his reach, himself not excepted; and his besom of destruction in the uncontrolled hands of a generation has swept into the sea soil fertility which only centuries of life could accumulate, and yet this fertility is the substratum of all that is living. It must be recognized that the phosphate deposits which we are beginning to return to our fields are but measures of fertility lost from older soils, and indices of processes

Fig. 108.—Type of conveyance extensively used in Japan for the removal of city and village waste. Such carts are even more frequently drawn by men than by cattle or horses, and tightly covered casks supported on saddles are borne on the backs of both cattle and horses, while men carry pails long distances on their shoulders, using the carrying pole.

still in progress. The rivers of North America are estimated to carry to the sea more than 500 tons of phosphorus with each cubic mile of water. To such loss modern civilization is adding that of hydraulic sewage disposal through which the waste of five hundred millions of people might be more than 194,300 tons of phosphorus annually, which could not be replaced by 1,295,000 tons of rock phosphate, 75 per cent pure. The Mongolian races, with a population now approaching the figure named; occupying an area little more than one-half that of the United States, tilling less than 800,000 square miles of land, and much of this during twenty, thirty or perhaps forty centuries; unable to avail themselves of mineral fertilizers, could not survive and tolerate such waste. Compelled to solve the problem of avoiding such wastes, and exercising the faculty which is characteristic of the race, they "cast down their buckets where they were", as

* A ship lost at sea for many days suddenly sighted a friendly vessel. From the mast of the unfortunate vessel was seen a signal, "Water, water; we die of thirst!" The answer from the friendly vessel at once came back, "Cast down your bucket where you are." A second time the signal, "Water, water; Send us water!" ran up from the distressed vessel, and was answered, "Cast down your bucket where you are." And a third and fourth signal for water was answered, "Cast down your bucket where you are." The captain of the distressed vessel, at last heeding the injunction, cast down his bucket, and it came up full of fresh sparkling water from the mouth of the Amazon river.

Not even in great cities like Canton, built in the meshes of tideswept rivers and canals; like Hankow on the banks of one of the largest rivers in the world; nor yet in modern Shanghai, Yokohama or Tokyo, is such waste permitted. To them such a practice has meant race suicide and they have resisted the temptation so long that it has ceased to exist.

Dr. Arthur Stanley, Health officer of the city of Shanghai, in his annual report for 1899, considering this subject as a municipal problem, wrote:

"Regarding the bearing on the sanitation of Shanghai of the relationship between Eastern and Western hygiene, it may be said, that if prolonged national life is indicative of sound sanitation, the Chinese are a race worthy of study by all who concern themselves with Public Health. Even without the returns of a Registrar-General it is evident that in China the birth rate must very considerably exceed the death rate, and have done so in an average way during the three or four thousand years that the Chinese nation has existed. Chinese hygiene, when compared with medieval English, appears to advantage. The main problem

---

*Booker T. Washington, Atlanta address.

of sanitation is to cleanse the dwelling day by day, and if this can be done at a profit so much the better. While the ultra-civilized Western elaborates destructors for burning garbage at a financial loss and turns sewage into the sea, the Chinaman uses both for manure. He wastes nothing while the sacred duty of agriculture is uppermost in his mind. And in reality recent bacterial work has shown that faecal matter and house refuse are best destroyed by returning them to clean soil, where natural purification takes place. The question of destroying garbage can, I think, under present conditions in Shanghai, be answered in a decided negative. While to adopt the water-carriage system for sewage and turn it into the river, whence the water supply is derived, would be an act of sanitary suicide. It is best, therefore, to make use of what is good in Chinese hygiene, which demands respect, being, as it is, the product of an evolution extending from more than a thousand years before the Christian era."

Fig. 109.—Receptacles for human waste.

The storage of such waste in China is largely in stoneware receptacles such as are seen in Fig. 109, which are hard-burned, glazed terra-cotta urns, having capacities ranging from 500 to 1000 pounds. Japan more often uses sheltered cement-lined pits such as are seen in Fig. 110.

In the three countries the carrying to the fields is often

est in some form of pail, as seen in Fig. 111, a pair of
which are borne swinging from the carrying pole.   In
applying the liquid to the field or garden the long handle
dipper is used, seen in Fig. 112.

We are beginning to husband with some economy the
waste from our domestic animals but in this we do not
approach that of China, Korea and Japan.   People in
China regularly search for and collect droppings along the
country and caravan roads.   Repeatedly, when walking

Fig. 110.—Japanese sheltered cement-lined storage pits for liquid manure.

through city streets, we observed such materials quickly
and apparently eagerly gathered, to be carefully stored
under conditions which ensure small loss from either leaching or unfavorable fermentation.   In some mulberry
orchards visited the earth had been carefully hoed back
about the trunks of trees to a depth of three or four inches
from a circle having a diameter of six to eight feet, and
upon these areas were placed the droppings of silkworms,
the moulted skins, together with the bits of leaves and
stem left after feeding.   Some disposition of such waste
must be made.   They return at once to the orchard all

but the silk produced from the leaves; unnecessary loss
is thus avoided and the material enters at once the service
of forcing the next crop of leaves.

On the farm of Mrs. Wu, near Kashing, while studying
the operation of two irrigation pumps driven by two cows,
lifting water to flood her twenty-five acres of rice field
preparatory to transplanting, we were surprised to observe

Fig. 111.—Six carrying pails such as are used in distributing liquid manure to
the fields.

that one of the duties of the lad who had charge of the
animals was to use a six-quart wooden dipper with a bamboo handle six feet long to collect all excreta, before they
fell upon the ground, and transfer them to a receptacle
provided for the purpose. There came a flash of resent-
ment that such a task was set for the lad, for we were
only beginning to realize to what lengths the practice of
economy may go, but there was nothing irksome suggested
in the boy's face. He performed the duty as a matter of
course and as we thought it through there was no reason
why it should have been otherwise. In fact, the only

right course was being taken. Conditions would have been worse if the collection had not been made. It made possible more rice. Character of substantial quality was building in the lad which meant thrift in the growing man and continued life for the nation.

We have adverted to the very small number of flies observed anywhere in the course of our travel, but its

Fig. 112.—Applying of liquid manure from carrying pails, using the long-handle dipper.

significance we did not realize until near the end of our stay. Indeed, for some reason, flies were more in evidence during the first two days on the steamship, out from Yokohama on our return trip to America, than at any time before on our journey. It is to be expected that the eternal vigilance which seizes every waste, once it has become such, putting it in places of usefulness, must contribute much toward the destruction of breeding places, and it may be these nations have been mindful of the wholesomeness of their practice and that many phases of the evolution of their waste disposal system have been dictated by and held fast to through a clear conception of sanitary needs.

Much intelligence and the highest skill are exhibited by these old-world farmers in the use of their wastes. In Fig. 113 is one of many examples which might be cited. The man walking down the row with his manure pails swinging from his shoulders informed us on his return that in his household there were twenty to be fed; that from this garden of half an acre of land he usually sold a product bringing in $400, Mexican,—$172, gold. The crop was cucumbers in groups of two rows thirty inches apart and twenty-four inches between the groups. The plants were eight to ten inches apart in the row. He had just marketed the last of a crop of greens which occupied the space between the rows of cucumbers seen under the strong, durable, light and very readily removable trellises. On May 28 the vines were beginning to run, so not a minute had been lost in the change of crop. On the contrary this man had added a month to his growing season by over-lapping his crops, and the trellises enabled him to feed more plants of this type than there was room for vines on the ground. With ingenuity and much labor he had made his half acre for cucumbers equivalent to more than two. He had removed the vines entirely from the ground; had provided a travel space two feet wide, down which he was walking, and he had made it possible to work about the roots of every plant for the purpose of hoeing and feeding. Four acres of cucumbers handled by American field methods would not yield more than this man's one, and he grows besides two other crops the same season. The difference is not so much in activity of muscle as it is in alertness and efficiency of the grey matter of the brain. He sees and treats each plant individually, he loosens the ground so that his liquid manure drops immediately beneath the surface within reach of the active roots. If the rainfall has been scanty and the soil is dry he may use ten of water to two of night soil, not to supply water but to make

Fig. 113.—Where the yield is the product of brain, brawn and utilized waste.

certain sufficiently deep penetration. If the weather is rainy and the soil over wet, the food is applied more concentrated, not to lighten the burden but to avoid waste by leaching and over saturation. While ever crowding growth he never overfeeds. Forethought, after-thought and the mind focussed on the work in hand are characteristic of these people. We do not recall to have seen a man smoking while at work. They enjoy smoking, but prefer to do this also with the attention undivided and thus get more for their money.

On another date earlier in May we were walking in the fields without an interpreter. For half an hour we stood watching an old gardener fitting the soil with his spading hoe in the manner seen in Fig. 26, where the graves of his ancestors occupy a part of the land. Angleworms were extremely numerous, as large around as an ordinary lead pencil and, when not extended, two-thirds as long, decidedly greenish in color. Nearly every stroke of the spade exposed two to five of these worms but so far as we observed, and we watched the man closely, pulverizing the soil, he neither injured nor left uncovered a single worm. While he seemed to make no effort to avoid injuring them or to cover them with earth, and while we could not talk with him, we are convinced that his action was continually guarded against injuring the worms. They certainly were subsoiling his garden deeply and making possible a freer circulation of air far below the surface. Their great abundance proved a high content of organic matter present in the soil and, as the worms ate their way through it, passing the soil through their bodies, the yearly volume of work done by them was very great. In the fields flooded preparatory to fitting them for rice these worms are forced to the surface in enormous numbers and large flocks of ducks are taken to such fields to feed upon them.

In another field a crop of barley was nearing maturity. An adjacent strip of land was to be fitted and planted. The leaning barley heads were in the way. Not one must be lost and every inch of ground must be put to use. The grain along the margin, for a breadth of sixteen inches, had been gathered into handfuls and skillfully tied, each with an unpulled barley stem, without breaking the straw, thus permitting even the grains in that head to fill and be gathered with the rest, while the tying set all straws well aslant, out of the way, and permitted the last inch of naked ground to be fitted without injuring the grain.

In still another instance a man was growing Irish potatoes to market when yet small. He had enriched his soil; he would apply water if the rains were not timely and sufficient, and had fed the plants. He had planted in rows only twelve to fourteen inches apart with a hill every eight inches in the row. The vines stood strong, straight, fourteen inches high and as even as a trimmed hedge. The leaves and stems were turgid, the deepest green and as prime and glossy as a prize steer. So close were the plants that there was leaf surface to intercept the sunshine falling on every square inch of the patch. There were no potato beetles and we saw no signs of injury but the gardener was scanning the patch with the eye of a robin. He spied the slightest first drooping of leaves in a stem; went after the difficulty and brought and placed in our hand a cut-worm, a young tuber the size of a marble and a stem cut half off, which he was willing to sacrifice because of our evident interest. But the two friends who had met were held apart by the babel of tongues.

Nothing is costing the world more; has made so many enemies, and has so much hindered the forming of friendships as the inability to fully understand; hence the dove that brings world peace must fly on the wings of a common language, and the bright star in the east is world commerce, rising on rapidly developing railway and steamship lines, heralded and directed by electric communication. With world commerce must come mutual confidence and friendship requiring a full understanding and therefore a common tongue. Then world peace will be permanently assured. It is coming inevitably and faster than we think. Once this desired end is seriously sought, the carrying of three generations of children through the public schools where the world language is taught together with the mother tongue, and the passing of the parents and grandparents, would effect the change.

The important point regarding these Far East people, to which attention should be directed, is that effective thinking, clear and strong, prevails among the farmers who have fed and are still feeding the dense populations from the products of their limited areas. This is further indicated in the universal and extensive use of plant ashes derived from fuel grown upon cultivated fields and upon the adjacent hill and mountain lands.

We were unable to secure exact data regarding the amount of fuel burned annually in these countries, and of ashes used as fertilizer, but a cord of dry oak wood weighs about 3500 pounds, and the weight of fuel used in the home and in manufactures must exceed that of two cords per household. Japan has an average of 5.563 people per family. If we allow but 1300 pounds of fuel per capita, Japan's consumption would be 31,200,000 tons. In view of the fact that a very large share of the fuel used in these countries is either agricultural plant stems, with an average ash content of 5 per cent, or the twigs and even leaves of trees, as in the case of pine bough fuel, 4.5 per cent of ash may be taken as a fair estimate. On this basis, and with a content of phosphorus equal to .5 per cent, and of potassium equal to 5 per cent, the fuel ash for Japan would amount to 1,404,000 tons annually, carrying 7020 tons of phosphorus and 70,200 tons of potassium, together with more than 400,000 tons of limestone, which is returned annually to less than 21,321 square miles of cultivated land.

In China, with her more than four hundred millions of people, a similar rate of fuel consumption would make the phosphorus and potassium returned to her fields more than eight times the amounts computed for Japan. On the basis of these statements Japan's annual saving of phosphorus from the waste of her fuel would be equivalent to more than 46,800 tons of rock phosphate having a purity of 75 per cent, or in the neighborhood of seven pounds per acre. If this amount, even with the potash and limestone added, appears like a trifling addition of fertility it is important for Americans to remember that even if this is so, these people have felt compelled to make the saving.

Fig. 114.—Japanese farmer tramping green herbage for fertilizer into the water and mud between rows of rice.

In the matter of returning soluble potassium to the cultivated fields Japan would be applying with her ashes the equivalent of no less than 156,600 tons of pure potassium sulphate, equal to 23 pounds per acre; while the lime carbonate so applied annually would be some 62 pounds per acre.

In addition to the forest lands, which have long been made to contribute plant food to the cultivated fields through fuel ashes, there are large areas which contribute green manure and compost material. These are chiefly hill lands, aggregating some twenty per cent of the cultivated fields, which bear mostly herbaceous growth. Some 2,552,741 acres of these lands may be cut over three times each season, yielding, in 1903, an average of 7980 pounds per acre. The first cutting of this hill herbage is mainly used on the rice fields as green manure, it being tramped into the mud between the rows after the manner seen in Fig. 114.

This man had been with basket and sickle to gather green herbage wherever he could and had brought it to his rice paddy. The day in July was extremely sultry. We came upon him wading in the water half way to his knees, carefully laying the herbage he had gathered between alternate rows of his rice, one handful in a place, with tips overlapping. This done he took the attitude seen in the illustration and, gathering the materials into a compact bunch, pressed it beneath the surface with his foot. The two hands smoothed the soft mud over the grass and righted the disturbed spears of rice in the two adjacent hills. Thus, foot following foot, one bare length ahead, the succeeding bunches of herbage were submerged until the last had been reached, following between alternate rows only a foot apart, there being a hill every nine to ten inches in the row and the hands grasping and being drawn over every one in the paddy.

He was renting the land, paying therefor forty kan of rice per tan, and his usual yield was eighty kan. This is forty-four bushels of sixty pounds per acre. In unfavorable seasons his yield might be less but still his rent would be forty kan per tan unless it was clear that he had done all that could reasonably be expected of him in securing the crop. It is difficult for Americans to understand how it is possible for the will of man, even when spurred by the love of home and family, to hold flesh to tasks like these.

The second and third cuttings of herbage from the *genya* lands in Japan are used for the preparation of compost applied on the dry-land fields in the fall or in the spring of the following season. Some of these lands are pastured,

14

but approximately 10,185,500 tons of green herbage grown
and gathered from the hills contributes much of its organic
matter and all of its ash to enrich the cultivated fields.
Such wild growth areas in Japan are the commons of the
near by villages, to which the people are freely admitted
for the purpose of cutting the herbage. A fixed time may

Fig. 115.—Father and children returning from *genya* lands with herbage for
use as green manure or for making compost. The daughter carries the tea
kettle to supply their safe, sanitary drink.

be set for cutting and a limit placed upon the amount
which may be carried away, which is done in the manner
seen in Fig. 115. It is well recognized by the people that
this constant cutting and removal of growth from the hill
lands, with no return, depletes the soils and reduces the
amount of green herbage they are able to secure.

Through the kindness of Dr. Daikuhara of the Imperial
Agricultural Experiment Station at Tokyo we are able to

give the average composition of the green leaves and young stems of five of the most common wild species of plants cut for green manure in June. In each 1000 pounds the amount of water is 562.18 pounds; of organic matter, 382.68 pounds; of ash, 55.14 pounds; nitrogen, 4.78 pounds; potassium, 2.407 pounds, and phosphorus, .34 pound. On the basis of this composition and an aggregate yield of 10,185,500 tons, there would be annually applied to the cultivated fields 3463 tons of phosphorus and 24,516 tons of potassium derived from the *genya* lands.

In addition to this the run-off from both the mountain and the *genya* lands is largely used upon the rice fields, more than sixteen inches of water being applied annually to them in some prefectures. If such waters have the composition of river waters in North America, twelve inches of water applied to the rice fields of the three main islands would contribute no less than 1200 tons of phosphorus and 19,000 tons of potassium annually.

Dr. Kawaguchi, of the National Department of Agriculture and Commerce, informed us that in 1908 Japanese farmers prepared and applied to their fields 22,812,787 tons of compost manufactured from the wastes of cattle, horses, swine and poultry, combined with herbage, straw and other similar wastes and with soil, sod or mud from ditches and canals. The amount of this compost is sufficient to apply 1.78 tons per acre of cultivated land of the southern three main islands.

From data obtained at the Nara Experiment Station, the composition of compost as there prepared shows it to contain, in each 2000 pounds, 550 pounds of organic matter; 15.6 pounds of nitrogen; 8.3 pounds of potassium, and 5.24 pounds of phosphorus. On this basis 22,800,000 tons of compost will carry 59,700 tons of phosphorus and 94,600 tons of potassium. The construction of compost houses is illustrated in Fig. 116, reproduced from a large circular sent to farmers from the Nara Experiment Station, and an exterior of one at the Nara Station is given in Fig. 117.

This compost house is designed to serve two and a half acres. Its floor is twelve by eighteen feet, rendered watertight by a mixture of clay, lime and sand. The walls are of earth, one foot thick, and the roof is thatched with straw. Its capacity is sixteen to twenty tons, having a

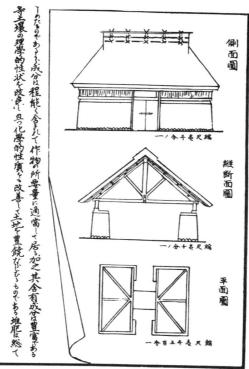

Fig. 116.—Section of chart issued by the Nara Experiment Station, illustrating construction of compost house; upper section shows elevation; middle portion is a cross section and the lower shows floor plan.

cash value of 60 yen, or $30. In preparing the stack, materials are brought daily and spread over one side of the compost floor until the pile has attained a hight of five feet. After one foot in depth has been laid and firmed, 1.2 inches of soil or mud is spread over the surface and the process repeated until full hight has been attained. Water is added sufficient to keep the whole saturated and to maintain the temperature below that of the body. After the compost stacks have been completed they are permitted to stand five weeks in summer, seven weeks in winter, when they are forked over and transferred to the opposite side of the house.

If we state in round numbers the total nitrogen, phosphorus and potassium thus far enumerated which Japanese farmers apply or return annually to their twenty or

Fig. 117.—Exterior view of compost house at Nara Experiment Station.

twenty-one thousand square miles of cultivated fields, the case stands 385,214 tons of nitrogen, 91,656 tons of phosphorus and 255,778 tons of potassium. These values are only approximations and do not include the large volume and variety of fertilizers prepared from fish, which have long been used. Neither do they include the very large amount of nitrogen derived directly from the atmosphere through their long, extensive and persistent cultivation of soy beans and other legumes. Indeed, from 1903 to 1906 the average area of paddy field upon which was grown a second crop of green manure in the form of some legume was 6.8 per cent of the total area of such fields aggregating 11,000 square miles. In 1906 over 18 per cent of the upland fields also produced some leguminous crop, these fields aggregating between 9,000 and 10,000 square miles.

While the values which have been given above, expressing the sum total of nitrogen, phosphorus and potassium applied annually to the cultivated fields of Japan may be somewhat too high for some of the sources named, there is little doubt that Japanese farmers apply to their fields more of these three plant food elements annually than has been computed. The amounts which have been given are sufficient to provide annually, for each acre of the 21,321 square miles of cultivated land, an application of not less than 56 pounds of nitrogen, 13 pounds of phosphorus and 37 pounds of potassium. Or, if we omit the large northern island of Hokkaido, still new in its agriculture and lacking the intensive practices of the older farm land, the quantities are sufficient for a mean application of 60, 14 and 40 pounds respectively of nitrogen, phosphorus and potassium per acre, and yet the maturing of 1000 pounds of wheat crop, covering grain and straw as water-free substance, removes from the soil but 13.9 pounds of nitrogen, 2.3 pounds of phosphorus and 8.4 pounds of potassium, from which it may be computed that the 60 pounds of nitrogen added is sufficient for a crop yielding 31 bushels of wheat; the phosphorus is sufficient for a crop of 44 bushels, and the potassium for a crop of 35 bushels per acre.

Dr. Hopkins, in his recent valuable work on "Soil Fertility and Permanent Agriculture" gives, on page 154, a table from which we abstract the following data:

APPROXIMATE AMOUNTS OF NITROGEN, PHOSPHORUS AND POTASSIUM
REMOVABLE PER ACRE ANNUALLY BY

	Nitrogen, pounds.	Phosphorus, pounds.	Potassium, pounds.
100 bush. crop of corn	148	23	71
100 bush. crop of oats	97	16	68
50 bush. crop of wheat	96	16	58
25 bush. crop of soy beans	159	21	73
100 bush. crop of rice	155	18	95
3 ton crop of timothy hay	72	9	71
4 ton crop of clover hay	160	20	120
3 ton crop of cow pea hay	130	14	98
8 ton crop of alfalfa hay	400	36	192
7,000 lb. crop of cotton	168	29.4	82
400 bush. crop of potatoes	84	17.3	120
20 ton crop of sugar beets	100	18	157
Annually applied in Japan, more than	60	14	40

We have inserted in this table, for comparison, the crop of rice, and have increased the crop of potatoes from three hundred bushels to four hundred bushels per acre, because such a yield, like all of those named, is quite practicable under good management and favorable seasons, notwithstanding the fact that much smaller yields are generally attained through lack of sufficient plant food or water. From this table, assuming that a crop of matured grain contains 11 per cent of water and the straw 15 per cent, while potatoes contain 79 per cent and beets 87 per cent, the amounts of the three plant food elements removable annually by 1000 pounds of crop have been calculated and stated in the next table.

APPROXIMATE AMOUNTS OF NITROGEN, PHOSPHORUS AND POTASSIUM
REMOVABLE ANNUALLY PER 1,0000 POUNDS OF DRY CROP SUBSTANCE.

	Nitrogen, pounds.	Phosphorus, pounds.	Potassium, pounds.
Cereals.			
Wheat	13.873	2.312	8.382
Oats	13.666	2.254	9.580
Corn	13.719	2.149	6.676
Legumes.			
Soy beans	30.807	4.070	14.147
Cow peas	25.490	2.745	19.216
Clover	23.529	2.941	17.647
Alfalfa	29.411	2.647	14.118
Roots.			
Beets	19.213	3.462	30.192
Potatoes	15.556	3.210	22.222
Grass.			
Timothy	14.117	1.765	13.922
Rice	9.949	1.129	6.089

From the amounts of nitrogen, phosphorus and potassium applied annually to the cultivated fields of Japan and from the data in these two tables it may be readily seen that these people are now and probably long have been applying quite as much of these three plant food elements to their fields with each planting as are removed with the crop, and if this is true in Japan it must also be true in China. Moreover there is nothing in American agricultural practice which indicates that we shall not ultimately be compelled to do likewise.

X.

IN THE SHANTUNG PROVINCE.

On May 15th we left Shanghai by one of the coastwise steamers for Tsingtao, some three hundred miles farther north, in the Shantung Province, our object being to keep in touch with methods of tillage and fertilization, corresponding phases of which would occur later in the season there.

The Shantung province is in the latitude of North Carolina and Kentucky, or lies between that of San Francisco and Los Angeles. It has an area of nearly 56,000 square miles, about that of Wisconsin. Less than one-half of this area is cultivated land yet it is at the present time supporting a population exceeding 38,000,000 of people. New York state has today less than ten millions and more than half of these are in New York city.

It was in this province that Confucius was born 2461 years ago, and that Mencius, his disciple, lived. Here, too, seventeen hundred years before Confucius' time, after one of the great floods of the Yellow river, 2297 B. C., and more than 4100 years ago, the Great Yu was appointed "Superintendent of Public Works" and entrusted with draining off the flood waters and canalizing the rivers.

Here also was the beginning of the Boxer uprising. Tsingtao sits at the entrance of Kiaochow Bay. Following the war of Japan with China this was seized by Germany, November 14, 1897, nominally to indemnify for the murder of two German missionaries which had occurred in Shantung, and March 6th, 1898, this bay, to the high water line, its islands and a "Sphere of Influence" extending thirty miles in all directions from the boundary, together with Tsingtao, was leased to Germany for ninety-nine years. Russia demanded and secured a lease of Port Arthur at the same time. Great Britain obtained a similar lease of Weihaiwei in Shantung, while to France Kwang-chow-wan in southern China, was leased. But the "encroachments" of European powers did not stop with these leases and during the latter part of 1898 the "Policy of Spheres of Influence" culminated in the international rivalry for railway concessions and mining. These greatly alarmed China and uprisings broke out very naturally first in Shantung, among the people nearest of kin to the founders of the Empire. As might have been expected of a patriotic, even though naturally peaceful people, they determined to defend their country against such encroachments and the Boxer troubles followed.

Tsingtao has a deep, commodious harbor always free from ice and Germany is constructing here very extensive and substantial harbor improvements which will be of lasting benefit to the province and the Empire. A pier four miles in length encloses the inner wharf, and a second wharf is nearing completion. Germany is also maintaining a meteorological observatory here and has established a large, comprehensive Forest Garden, under excellent management, which is showing remarkable developments for so short a time.

Our steamer entered the harbor during the night and, on going ashore, we soon found that only Chinese and German were generally spoken; but through the kind assistance of Rev. W. H. Scott, of the American Presbyterian Mission, an interpreter promised to call at my hotel in the evening, although he failed to appear. The afternoon was spent at the Forest Garden and on the reforestation tract, which are under the supervision of Mr. Haas. The Forest Garden covers two hundred and seventy acres and the reforestation tract three thousand acres more. In the garden a great variety of forest and fruit trees and small fruits are being tried out with high promise of the most valuable results.

It was in the steep hills about Tsingtao that we first saw at close range serious soil erosion in China; and the returning of forest growth on hills nearly devoid of soil

Fig. 118.—Granite hill destitute of soil, rapidly falling into decay. Reforestation area, Tsingtao, Stantung.

was here remarkable, in view of the long dry seasons which prevail from November to June, and Fig. 118 shows how destitute of soil the crests of granite hills may become and yet how the coming back of the forest growth may hasten as soon as it is no longer cut away. The rock going into decay, where this view was taken, is an extremely coarse crystalline granite, as may be seen in contrast with the watch, and it is falling into decay at a marvelous rate.

Fig 119.—Almost soil-free granite surface on which young forest growth, largely pine, is developing. Reforestation tract, Tsingtao, Shantung.

Disintegration has penetrated the rock far below the surface and the large crystals are held together with but little more tenacity than prevails in a bed of gravel. Moisture and even roots penetrate it deeply and readily and the crystals fall apart with thrusts of the knife blade, the rock crumbling with the greatest freedom. Roadways have been extensively carved along the sides of the hills with the aid of only pick and shovel. Close examination of the rock shows that layers of sediment exist between the crystal

Fig. 120.—Forest and herbaceous growth coming back over such soil conditions as are seen in Figs. 118 and 119. Reforestation tract, Tsingtao, Shantung.

faces, either washed down by percolating rain or formed through decomposition of the crystals in place. The next illustration, Fig. 119, shows how large the growth on such soils may be, and in Fig. 120 the vegetation and forest growth are seen coming back, closely covering just such soil surfaces and rock structure as are indicated in Figs. 118 and 119.

These views are taken on the reforestation tract at Tsingtao but most of the growth is volunteer, standing now protected by the German government in their effort to see what may be possible under careful supervision.

The loads of pine bough fuel represented in Fig. 80 were gathered from such hills and from such forest growth as are here represented, but on lands more distant from the city. But Tsingtao, with its forty thousand Chinese, and Kiaochow across the bay, with its one hundred and twenty thousand more, and other villages dotting the narrow plains, maintain a very great demand for such growth on the hill lands. The wonder is that forest growth has persisted at all and has contributed so much in the way of fuel.

Fig. 121.—Close view of the wild yellow Shantung rose cultivated in the Forest Garden at Tsingtao and very effective for parks and pleasure drives.

Growing in the Forest Garden was a most beautiful wild yellow rose, native to Shantung, being used for landscape effect in the parking, and it ought to be widely introduced into other countries wherever it will thrive. It was growing as heavy borders and massive clumps six to eight feet high, giving a most wonderful effect, with its brilliant, dense cloud of the richest yellow bloom. The blossoms are single, fully as large as the Rosa rugosa, with the tips of the petals shading into the most dainty light straw yellow, while the center is a deep orange, the contrast being sufficient to show in the photograph from which Fig. 121 was prepared. Another beautiful and striking feature of this rose is the clustering of the blossoms in one-sided wreath-like sprays, sometimes twelve to eighteen inches long, the flowers standing close enough to even overlap.

The interpreter engaged for us failed to appear as per agreement so the next morning we took the early train for Tsinan to obtain a general view of the country and to note the places most favorable as points for field study. We had resolved also to make an effort to secure an interpreter through the American Presbyterian College at Tsinan. Leaving Tsingtao, the train skirts around the Kiaochow bay for a distance of nearly fifty miles, where we pass the city of the same name with its population of 120,000, which had an import and export trade in 1905 valued at over $24,000,000. At Sochen we passed through a coal mining district where coal was being brought to the cars in baskets carried by men. The coal on the loaded open cars was sprinkled with whitewash, serving as a seal to safe-guard against stealing during transit, making it so that none could be removed without the fact being revealed by breaking the seal. This practice is general in China and is applied to many commodities handled in bulk. We saw baskets of milled rice carried by coolies sealed with a pattern laid over the surface by sprinkling some colored powder upon it. Cut stone, corded for the market, was whitewashed in the same manner as the coal.

As we were approaching Weihsien, another city of 100,000 people, we identified one of the deeply depressed, centuries-old roadways, worn eight to ten feet deep, by chancing to see half a dozen teams passing along it as the train crossed. We had passed several and were puzzling to account for such peculiar erosion. The teams gave the explanation and thus connected our earlier reading with the concrete. Along these deep-cut roadways caravans may pass, winding through the fields, entirely unobserved unless one chances to be close along the line or the movement is discovered by clouds of dust, one of the methods that has produced them, and we would not be surprised if gathering manure from them has played a large part also.

Weihsien is near one of the great commercial highways of China and in the center of one of the coal mining regions of the province. Still further along towards Tsinan we passed Tsingchowfu, another of the large cities of the province, with 150,000 population. All day we rode through fields of wheat, always planted in rows, and in hills in the row east of Kaumi, but in single or double continuous drills westward from here to Tsinan. Thousands of wells used for irrigation, of the type seen in Fig. 123, were passed during the day, many of them recently dug to supply water for the barley suffering from the severe drought which was threatening the crop at the time.

It was 6:30 P. M. before our train pulled into the station at Tsinan; 7:30 when we had finished supper and engaged a ricksha to take us to the American Presbyterian College in quest of an interpreter. We could not speak Chinese, the ricksha boy could neither speak nor understand a word of English, but the hotel proprietor had instructed him where to go. We plunged into the narrow streets of a great Chinese city, the boy running wherever he could, walking where he must on account of the density of the crowds or the roughness of the stone paving. We had turned many corners, crossed bridges and passed through tunneled archways in sections of the massive city walls, until it was getting dusk and the ricksha man purchased and lighted a lantern. We were to reach the college in thirty minutes but had been out a full hour. A little later the boy drew up to and held conference with a policeman. The curious of the street gathered about and it dawned upon us that we were lost in the night in the narrow streets of a Chinese city of a hundred thousand people. To go further would be useless for the gates of the mission compound would be locked. We could only indicate by motions our desire to return, but these were not understood. On the train a thoughtful, kindly old

German had recognized a stranger in a foreign land and volunteered useful information, cutting from his daily paper an advertisement describing a good hotel. This gave the name of the hotel in German, English and in Chinese characters. We handed this to the policeman, pointing to the name of the hotel, indicating by motions the desire to return, but apparently he was unable to read in either language and seemed to think we were assuming to direct the way to the college. A man and boy in the crowd apparently volunteered to act as escort for us. The throng parted and we left them, turned more corners into more unlighted narrow alleyways, one of which was too difficult to permit us to ride. The escorts, if such they were, finally left us, but the dark alley led on until it terminated at the blank face, probably of some other portion of the massive city wall we had thrice threaded through lighted tunnels. Here the ricksha boy stopped and turned about but the light from his lantern was too feeble to permit reading the workings of his mind through his face, and our tongues were both utterly useless in this emergency, so we motioned for him to turn back and by some route we reached the hotel at 11 P. M.

We abandoned the effort to visit the college, for the purpose of securing an interpreter, and took the early train back to Tsingtao, reaching there in time to secure the very satisfactory service of Mr. Chu Wei Yung, through the further kind offices of Mr. Scott. We had been twice over the road between the two cities, obtaining a general idea of the country and of the crops and field operations at this season. The next morning we took an early train to Tsangkau and were ready to walk through the fields and to talk with the last generations of more than forty unbroken centuries of farmers who, with brain and brawn, have successfully and continuously sustained large families on small areas without impoverishing their soil. The next illustration is from a photograph taken in one of these fields. We astonished the old farmer by asking the privilege of holding his plow through one round in his little field, but he granted the privilege readily. Our furrow was not as well turned as his, nor as well as we could have done with a two-handled Oliver or John Deere, but it was better than the old man had expected and won his respect.

This plow had a good steel point, as a separate, blunt, V-shaped piece, and a moldboard of cast steel with a good twist which turned the soil well. The standard and sole were of wood and at the end of the beam was a block for gauging the depth of furrow. The cost of this plow, to

Fig. 122.—A Shantung plow, simple but effective.

the farmer, was $2.15, gold, and when the day's work is done it is taken home on the shoulders, even though the distance may be a mile or more, and carefully housed. Chinese history states that the plow was invented by Shennung, who lived 2737–2697 B. C. and "taught the art of agriculture and the medical use of herbs". He is honored as the "God of Agriculture and Medicine."

Through my interpreter we learned that there were twelve in this man's family, which he maintained on fifteen mow of land, or 2.5 acres, together with his team, consisting of a cow and small donkey, besides feeding two pigs.

15

This is at the rate of 192 people, 16 cows, 16 donkeys and 32 pigs on a forty-acre farm; and of a population density equivalent to 3072 people, 256 cows, 256 donkeys and 512 swine per square mile of cultivated field.

On another small holding we talked with the farmer standing at the well in Fig. 27, where he was irrigating a little piece of barley 30 feet wide and 138 feet long. He owned and was cultivating but one and two-thirds acres of land and yet there were ten in his family and he kept one donkey and usually one pig. Here is a maintenance capacity at the rate of 240 people, 24 donkeys and 24 pigs on a forty-acre farm; and a population density of 3840 people, 384 donkeys and 384 pigs per square mile. His usual annual sales in good seasons were equivalent in value to $73, gold.

In both of these cases the crops grown were wheat, barley, large and small millet, sweet potatoes and soy beans or peanuts. Much straw braid is manufactured in the province by the women and children in their homes, and the cargo of the steamer on which we returned to Shanghai consisted almost entirely of shelled peanuts in gunny sacks and huge bales of straw braid destined for the manufacture of hats in Europe and America.

Shantung has only moderate rainfall, little more than 24 inches annually, and this fact has played an important part in determining the agricultural practices of these very old people. In Fig. 123 is a closer view than Fig. 27 of the farmer watering his little field of barley. The well had just been dug over eight feet deep, expressly and solely to water this one piece of grain once, after which it would be filled and the ground planted.

The season had been unusually dry, as had been the one before, and the people were fearing famine. Only 2.44 inches of rain had fallen at Tsingtao between the end of the preceding October and our visit, May 21st, and hundreds of such temporary wells had been or were being dug all along both sides of the two hundred and fifty miles of railway, and nearly all to be filled when the crop on

Fig. 123.—Temporary well and portable irrigation outfit, Shantung, China.

the ground was irrigated, to release the land for one to follow. The homes are in villages a mile or more apart and often the holdings or rentals are scattered, separated by considerable distances, hence easy portability is the key-note in the construction of this irrigating outfit. The bucket is very light, simply a woven basket waterproofed with a paste of bean flour. The windlass turns like a long spool on a single pin and the standard is a tripod with removable legs. Some wells we saw were sixteen or twenty feet deep and in these the water was raised by a cow walking straight away at the end of a rope.

The amount and distribution of rainfall in this province, as indicated by the mean of ten years' records at Tsingtao, obtained at the German Meteorological Observatory through the courtesy of Dr. B. Meyermanns, are given in the table in which the rainfall of Madison, Wisconsin, is inserted for comparison.

	Mean monthly rainfall.		Mean rainfall in 10 days.	
	Tsingtao, Inches.	Madison, Inches.	Tsingtao, Inches.	Madison, Inches.
January	.394	1.56	.131	.520
February	.240	1.50	.080	.500
March	.892	2.12	.297	.707
April	1.240	2.52	.413	.840
May	1.636	3.62	.545	1.207
June	2.702	4.10	.901	1.366
July	6.637	3.90	2.212	1.300
August	5.157	3.21	1.719	1.070
September	2.448	3.15	.816	1.050
October	2.258	2.42	.753	.807
November	.396	1.78	.132	.593
December	.682	1.77	.227	.590
Total	24.682	31.65		

While Shantung receives less than 25 inches of rain during the year, against Wisconsin's more than 31 inches, the rainfall during June, July and August in Shantung is nearly 14.5 inches, while Wisconsin receives but 11.2 inches. This greater summer rainfall, with persistent fertilization and intense management, in a warm latitude, are some of the elements permitting Shantung today to feed 38,247,900 people from an area equal to that upon which Wisconsin is yet feeding but 2,333,860. Must American agriculture ultimately feed sixteen people where it is now feeding but one? If so, correspondingly more intense and effective practices must follow, and we can neither know too well nor too early what these Old World people have been driven to do; how they have succeeded, and how we and they may improve upon their practices and lighten the human burdens by more fully utilizing physical forces and mechanical appliances.

As we passed on to other fields we found a mother and daughter transplanting sweet potatoes on carefully fitted

Fig. 124.—Strong erosion in Shantung, with wheat on remnants of tables.

ridges of nearly air-dry soil in a little field, the remnant of a table on a deeply eroded hillside, Fig. 124. The husband was bringing water for moistening the soil from a deep ravine a quarter of a mile distant, carrying it on his shoulder in two buckets, Fig. 125, across an intervening gulch. He had excavated four holes at intervals up the gulch and from these, with a broken gourd dipper mended with stitches, he filled his pails, bailing in succession from one to the other in regular rotation.

The daughter was transplanting. Holding the slip with its tip between thumb and fingers, a strong forward stroke plowed a furrow in the mellow, dry soil; then, with a backward movement and a downward thrust, planted the slip, firmed the soil about it, leaving a depression in which the mother poured about a pint of water from another gourd dipper. After this water had soaked away, dry earth was drawn about the slip and firmed and looser earth drawn over this, the only tools being the naked hands and dipper.

The father and mother were dressed in coarse garb but the daughter was neatly clad, with delicate hands decorated with rings and a bracelet. Neither of the women had bound

Fig. 125.—Getting water to transplant sweet potatoes. A Standard Oil can is balanced against China's ancient stone jar.

feet. There were ten in his family; and on adjacent similar areas they had small patches of wheat nearly ready for the harvest, all planted in hills, hoed, and in astonishingly vigorous condition considering the extreme drought which prevailed. The potatoes were being planted under these extreme conditions in anticipation of the rainy season which then was fully due. The summer before had been one of unusual drought, and famine was threatened. The government had recently issued an edict that no sheep should be sold from the province, fearing they might be needed for food. An old woman in one of the villages came out,

Fig. 126.—Two views of the same farmyard, showing a pile of prepared compost and the farm team.

as we walked through, and inquired of my interpreter if we had come to make it rain. Such was the stress under which we found these people.

One of the large farmers, owning ten acres, stated that his usual yield of wheat in good season was 160 catty per mow, equivalent to 21.3 bushels per acre. He was expecting the current season not more than one-half this amount. As a fertilizer he used a prepared earth compost which we shall describe later, mixing it with the grain and sowing in the hills with the seed, applying about 5333 pounds per acre, which he valued, in our currency, at $8.60, or $3.22 per ton. A pile of such prepared compost is seen in Fig. 126, ready to be transferred to the field. The views show with what cleanliness the yard is kept and with what care all animal waste is saved. The cow and donkey are the work team, such as was being used by the plowman referred to in Fig. 122. The mounds in the background of the lower view are graves; the fence behind the animals is made from the stems of the large millet, kaoliang, while that at the right of the donkey is made of earth, both indicative of the scarcity of lumber. The buildings, too, are thatched and their walls are of earth plastered with an earthen mortar worked up with chaff.

In another field a man plowing and fertilizing for sweet potatoes had brought to the field and laid down in piles the finely pulverized dry compost. The father was plowing; his son of sixteen years was following and scattering, from a basket, the pulverized dry compost in the bottom of the furrow. The next furrow covered the fertilizer, four turned together forming a ridge upon which the potatoes were to be planted after a second and older son had smoothed and fitted the crest with a heavy hand rake. The fertilizer was thus applied directly beneath the row, at the rate of 7400 pounds per acre, valued at $7.15, our currency, or $1.93 per ton.

We were astonished at the moist condition of the soil turned, which was such as to pack in the hand notwithstanding the extreme drought prevailing and the fact that standing water in the ground was more than eight feet below the surface. The field had been without crop and cultivated.

To the question, "What yield of sweet potatoes do you expect from this piece of land?" he replied, "About 4000 catty," which is 440 bushels of 56 pounds per acre. The usual market price was stated to be $1.00, Mexican, per one hundred catty, making the gross value of the crop $79.49, gold, per acre. His land was valued at $60, Mexican, per mow, or $154.80 per acre, gold.

My interpreter informed me that the average well-to-do farmers in this part of Shantung own from fifteen to twenty mow of land and this amount is quite ample to provide for eight people. Such farmers usually keep two cows, two donkeys and eight or ten pigs. The less well-to-do or small farmers own two to five mow and act as superintendents for the larger farmers. Taking the largest holding, of twenty mow per family of eight people, as a basis, the density per square mile would be 1536 people, and an area of farm land equal to the state of Wisconsin would have 86,000,000 people; 21,500,000 cows; 21,500,000 donkeys and 86,000,000 swine. These observations apply to one of the most productive sections of the province, but very large areas of land in the province are not cultivable and the last census showed the total population nearly one-half of this amount. It is clear, therefore, that either very effective agricultural methods are practiced or else extreme economy is exercised. Both are true.

On this day in the fields our interpreter procured his dinner at a farm house, bringing us four boiled eggs, for which he paid at the rate of 8.3 cents of our money, but his dinner was probably included in the price. The next table gives the prices for some articles obtained by inquiry at the Tsingtao market, May 23rd, 1909, reduced to our currency.

	Cents.
Old potatoes, per lb	2.18
New potatoes, per lb	2.87
Salted turnip. per lb	.86
Onions, per lb	4.10
Radishes, bunch of 10	1.29
String beans, per lb	11.46
Cucumbers, per lb	5.73
Pears, per lb	5.73
Apricots. per lb	8.60
Pork, fresh. per lb	10.33
Fish, per lb	5.73
Eggs, per dozen	5.16

The only items which are low compared with our own prices are salted turnips, radishes and eggs. Most of the articles listed were out of season for the locality and were imported for the foreigners, turnips, radishes, pork, fish and eggs being the exceptions. Prof. Ross informs us that he found eggs selling in Shensi at four for one cent of our money.

Our interpreter asked a compensation of one dollar, Mexican, or 43 cents, U. S. currency, per day, he furnishing his own meals. The usual wage for farm labor here was $8.60, per year, with board and lodging. We have referred to the wages paid by missionaries for domestic service. As servants the Chinese are considered efficient, faithful and trustworthy. It was the custom of Mr. and Mrs. League to intrust them with the purse for marketing, feeling that they could be depended upon for the closest bargaining. Commonly, when instructed to procure a certain article, if they found the price one or two cash higher than usual they would select a cheaper substitute. If questioned as to why instructions were not followed the reply would be "Too high, no can afford."

Mrs. League recited her experience with her cook regarding his use of our kitchen appliances. After fitting the kitchen with a modern range and cooking utensils, and working with him to familiarize him with their use, she was surprised, on going into the kitchen a few days later, to find that the old Chinese stove had been set on the range and the cooking being done with the usual Chinese furniture. When asked why he was not using the stove his reply was "Take too much fire." Nothing jars on the nerves of these people more than incurring of needless expense, extravagance in any form, or poor judgment in making purchases.

Daily we became more and more impressed by the evidence of the intense and incessant stress imposed by the dense populations of centuries, and how, under it, the laws of heredity have wrought upon the people, affecting constitution, habits and character. Even the cattle and sheep have not escaped its irresistable power. Many times in this province we saw men herding flocks of twenty to thirty sheep along the narrow unfenced pathways winding through the fields, and on the grave lands. The prevailing drought had left very little green to be had from these places and yet sheep were literally brushing their sides against fresh green wheat and barley, never molesting them. Time and again the flocks were stampeded into the grain by an approaching train, but immediately they returned to their places without taking a nibble. The voice of the shepherd and an occasional well aimed lump of earth only being required to bring them back to their uninviting pastures.

In Kiangsu and Chekiang provinces a line of half a dozen white goats were often seen feeding single file along the pathways, held by a cord like a string of beads, sometimes led by a child. Here, too, one of the most common sights was the water buffalo grazing unattended among the fields along the paths and canal banks, with crops all about. One of the most memorable shocks came to us in Chekiang, China, when we had fallen into a revery while gazing at the shifting landscape from the doorway of our low-down Chinese houseboat. Something in the sky and the vegetation along the canal bank had recalled the scenes of boyhood days and it seemed, as we looked aslant up the bank with its fringe of grass, that we were gliding along Whitewater creek through familiar meadows and that standing up would bring the old home in sight. That instant there glided into view, framed in the doorway and projected high against the tinted sky above the setting sun, a giant water buffalo standing motionless as a statue on the summit of a huge grave mound, lifted fully ten feet above the field. But in a flash this was replaced by a companion scene, and with all its beautiful setting, which had been as suddenly fixed on the memory fourteen years before in the far away Trossachs when our coach, hurriedly rounding a sharp turn in the hills, suddenly exposed a wild ox of Scotland similarly thrust against the sky from a small but isolated rocky summit, and then, outspeeding the wireless, recollection crossed two oceans and an intervening continent, bringing us back to China before a speed of five miles per hour could move the first picture across the narrow doorway.

It was through the fields about Tsangkow that the stalwart freighters referred to, Fig. 32, passed us on one of the paths leading from Kiaochow through unnumbered country villages, already eleven miles on their way with their wheelbarrows loaded with matches made in Japan. Many of the wheelbarrow men seen in Shanghai and other cities are from Shantung families, away for employment, expecting to return. During the harvest season, too, many of these people go west and north into Manchuria seeking employment, returning to their homes in winter.

Alexander Hosie, in his book on Manchuria, states that from Chefoo alone more than 20,000 Chinese laborers cross to Newchwang every spring by steamer, others finding their way there by junks or other means, so that after the harvest season 8,000 more return by steamer to Chefoo than left that way in the spring, from which he concludes that Shantung annually supplies Manchuria with agricultural labor to the extent of 30,000 men.

About the average condition of wheat in Shantung during this dry season, and nearing maturity, is seen in Fig. 127, standing rather more than three feet high, as indicated by our umbrella between the rows. Beyond the wheat and to the right, grave mounds serrate the sky line, no hills being in sight, for we were in the broad plain built up from the sea between the two mountain islands forming the highlands of Shantung.

On May 22nd we were in the fields north of Kiaochow, some sixty miles by rail west from Tsingtao, but within the neutral zone extending thirty miles back from the high water line of the bay of the same name. Here the Germans had built a broad macadam road after the best European type but over it were passing the vehicles of forty centuries seen in Figs. 128 and 129. It is doubtful if the resistance to travel experienced by these men on the better road was enough less than that on the old paths they had left to convince them that the cost of construction and maintenance would be worth while until vehicles and the price of labor change. It may appear strange that with a nation of so many millions and with so long a history, roads have persisted as little more than beaten foot-paths; but modern methods of transportation have remained phy-sical impossibilities to every people until the science of the

Fig. 127.—Field of wheat in Shantung, China, nearing maturity in a season of unusual drought.

last century opened the way. Throughout their history the burdens of these people have been carried largely on foot, mostly on the feet of men, and of single men wherever the load could be advantageously divided. Animals have been supplemental burden bearers but, as with the men, they have carried the load directly on their own feet, the mode least disturbed by inequalities of road surface.

For adaptability to the worst road conditions no vehicle equals the wheelbarrow, progressing by one wheel and two feet. No vehicle is used more in China, if the carrying

Figs. 128 and 129.—The vehicles of forty centuries on a modern road of German construction, Kiaochow, Shantung, China.

pole is excepted, and no wheelbarrow in the world permits so high an efficiency of human power as the Chinese, as must be clear from Figs. 32 and 61, where nearly the whole load is balanced on the axle of a high, massive wheel with broad tire. A shoulder band from the handles of the barrow relieves the strain on the hands and, when the load or the road is heavy, men or animals may aid in drawing, or even, when the wind is favorable, it is not unusual to hoist a sail to gain propelling power. It is only in northern China, and then in the more level portions, where there are few or no canals, that carts have been extensively used, but are more difficult to manage on bad roads. Most of the heavy carts, especially those in Manchuria, seen in Fig. 203, have the wheels framed rigidly to the axle which revolves with them, the bearing being in the bed of the cart. But new carts of modern type are being introduced.

In the extent of development and utilization of inland waterways no people have approached the Chinese. In the matter of land transportation they have clearly followed the line of least resistance for individual initiative, so characteristic of industrial China.

There are Government courier or postal roads which connect Peking with the most distant parts of the Empire, some twenty-one being usually enumerated. These, as far as practicable, take the shortest course, are often cut into the mountain sides and even pass through tunnels. In the plains regions these roads may be sixty to seventy-five feet wide, paved and occasionally bordered by rows of trees. In some cases, too, signal towers are erected at intervals of three miles and there are inns along the way, relay posts and stations for soldiers.

We have spoken of planting grain in rows and in hills in the row. In Fig. 130 is a field with the rows planted in pairs, the members being 16 inches apart, and together occupying 30 inches. The space between each pair is also 30 inches, making five feet in all. This makes frequent hoeing practicable, which is begun early in the spring and is repeated after every rain. It also makes it possible to feed the plants when they can utilize food to the best advantage and to repeat the feeding if desirable. Besides, the ground in the wider space may be fitted, fertilized and another crop planted before the first is removed. The hills alternate in the rows and are 24 to 26 inches from center to center.

The planting may be done by hand or with a drill such as that in Fig. 131, ingenious in the simple mechanism

Fig. 130.—Wheat planted in hills and in rows, the pairs of rows being 30 inches apart and the rows 16 inches, covering 5 feet.

which permits planting in hills. The husbandman had just returned from the field with the drill on his shoulder when we met at the door of his village home, where he explained to us the construction and operation of the drill and permitted the photograph to be taken, but turning his face aside, not wishing to represent a specific character, in the view. In the drill there was a heavy leaden weight swinging free from a point above the space between the openings leading to the respective drill feet. When planting, the operator rocks the drill from side to side, causing the weight to hang first over one and then over the other opening, thus securing alternation of hills in each pair of rows.

Counting the heads of wheat in the hill in a number of fields showed them ranging between 20 and 100, the distance between the rows and between the hills as stated above. There were always a larger number of stalks per hill where the water capacity of the soil was large, where the ground water was near the surface, and where the soil

Fig. 131.—Double row seed-drill, just returning from the fields to the village home.

was evidently of good quality. This may have been partly the result of stooling but we have little doubt that judgment was exercised in planting, sowing less seed on the lighter soils where less moisture was available. In the piece just referred to, in the illustration, an average hill contained 46 stalks and the number of kernels in a head varied between 20 and 30. Taking Richardson's estimate of 12,000 kernels of wheat to the pound, this field would yield about twelve bushels of wheat per acre this unusually dry season. Our interpreter, whose parents lived near Kaomi, four stations further west, stated that in 1901, one of their best seasons, farmers there secured yields as high as 875 catty per legal mow, which is at the rate of 116 bushels per acre. Such a yield on small areas highly fertilized and carefully tilled, when the rainfall is ample or where irrigation is practiced, is quite possible and in the Kiangsu province we observed individual small fields which would certainly approach close to this figure.

Further along in our journey of the day we came upon a field where three, one of them a boy of fourteen years, were hoeing and thinning millet and maize. In China, during the hot weather, the only garment worn by the men in the field, was their trousers, and the boy had found these unnecessary, although he slipped into them while we were talking with his father. The usual yield of maize was set at 420 to 480 catty per mow, and that of millet at 600 catty, or 60 to 68.5 bushels of maize and 96 bushels of millet, of fifty pounds, per acre, and the usual price would make the gross earnings $23.48 to $26.83 per acre for the maize, and $30.96, gold, for the millet.

It was evident when walking through these fields that the fall-sowed grain was standing the drought far better than the barley planted in the spring, quite likely because of the deeper and stronger development of root system made possible by the longer period of growth, and partly because the wheat had made much of its growth utilizing water that had fallen before the barley was planted and which would have been lost from the soil through percolation and surface evaporation. Farmers here are very particular to hoe their grain, beginning in the early spring, and always after rains, thoroughly appreciating the efficiency of earth mulches. Their hoe, seen in Fig. 132, is peculiarly well adapted to its purpose, the broad blade being so hung that it draws nearly parallel with the surface, cutting shallow and permitting the soil to drop practically upon the place from which it was loosened. These hoes are made in three parts; a wooden handle, a long, strong and heavy iron socket shank, and a blade of steel. The blade is detachable and different forms and sizes of blades may be used on the same shank. The mulch-producing blades may have a cutting edge thirteen inches long and a width of nine inches.

At short intervals on either hand, along the two hundred and fifty miles of railway between Tsingtao and Tsinan, were observed many piles of earth compost dis-

Fig. 132.—Method of using the broad, heavy hoe in producing surface mulch, as seen in Shantung, China.

tributed in the fields. One of these piles is seen in Fig. 133. They were sometimes on unplanted fields, in other cases they occurred among the growing crops soon to be harvested, or where another crop was to be planted between the rows of one already on the ground. Some of these piles were six feet high. All were built in cubical form with flat top and carefully plastered with a layer of earth mortar which sometimes cracked on drying, as seen in the illustration. The purpose of this careful shaping and plastering we did not learn although our interpreter stated it was to prevent the compost from being appropriated for use on adjacent fields. Such a finish would have the effect of a seal, showing if the pile had been disturbed, but we suspect other advantages are sought by the treatment, which involves so large an amount of labor.

The amount of this earth compost prepared and used

Fig. 133.—Carefully plastered earth compost stacked in the field awaiting distribution, Shantung, China.

annually in Shantung is large, as indicated by the cases cited, where more than five thousand pounds, in one instance, and seven thousand pounds in another, were applied per acre for one crop. When two or more crops are grown the same year on the same ground, each is fertilized, hence from three to six or more tons may be applied to each cultivated acre. The methods of preparing compost and of fertilizing in Kiangsu, Chekiang and Kwangtung provinces have been described. In this part of Shantung, in Chihli and north in Manchuria as far as Mukden, the methods are materially different and if possible even more laborious, but clearly rational and effective. Here nearly if not all fertilizer compost is prepared in the villages and carried to the fields, however distant these may be.

Rev. T. J. League very kindly accompanied us to Chengyang on the railway, from which we walked some two miles back to a prosperous rural village to see their methods of preparing this compost fertilizer. It was toward the close of the afternoon before we reached the village, and from all directions husbandmen were returning from the fields, some with hoes, some with plows, some with drills over their shoulders and others leading donkeys or cattle, and similar customs obtain in Japan, as seen in Fig. 134. These were mostly the younger men. When we reached the village streets the older men, all bareheaded, as were those returning from the fields, and usually with their queues tied about the crown, were visiting, enjoying their pipes of tobacco.

Opium is no longer used openly in China, unless it be permitted to some well along in years with the habit confirmed, and the growing of the poppy is prohibited. The penalties for violating the law are heavy and enforcement is said to be rigid and effective. For the first violation a fine is imposed. If convicted of a second violation the fine is heavier with imprisonment added to help the victim acquire self control, and a third conviction may bring the death penalty. The eradication of the opium scourge must prove a great blessing to China. But with the passing of this most formidable evil, for whose infliction upon China England was largely responsible, it is a great misfortune that through the pitiless efforts of the British-American Tobacco Company her people are rapidly becoming addicted to the western tobacco habit, selfish beyond excuse, filthy beyond measure, and unsanitary in its polluting and oxygen-destroying effect upon the air all are compelled to breathe. It has already become a greater and more inexcusable burden upon mankind than opium ever was.

China, with her already overtaxed fields, can ill afford

to give over an acre to the cultivation of this crop and she should prohibit the growing of tobacco as she has that of the poppy. Let her take the wise step now when she readily may, for all civilized nations will ultimately be compelled to adopt such a measure. The United States in 1902 had more than a million acres growing tobacco, and harvested 821,000,000 pounds of leaf. This leaf depleted those soils to the extent of more than twenty-

Fig. 134.—Home after the day's work, in Japan.

eight million pounds of nitrogen, twenty-nine million pounds of potassium and nearly two and a half million pounds of phosphorus, all so irrecoverably lost that even China, with her remarkable skill in saving and her infinite patience with little things, could not recover them for her soils. On a like area of field might as readily be grown twenty million bushels of wheat and if the twelve hundred million pounds of grain were all exported it would deplete the soil less than the tobacco crop in everything but phosphorus, and in this about the same. Used at home, China would return it all to one or another field.

The home consumption of tobacco in the United States averaged seven pounds per capita in 1902. A like consumption for China's four hundred millions would call for 2800 million pounds of leaf. If she grew it on her fields two million acres would not suffice. Her soils would be proportionately depleted and she would be short forty million bushels of wheat; but if China continues to import her tobacco the vast sum expended can neither fertilize her fields nor feed, clothe or educate her people, yet a like sum expended in the importation of wheat would feed her hungry and enrich her soils.

In the matter of conservation of national resources here is one of the greatest opportunities open to all civilized nations. What might not be done in the United States with a fund of $57,000,000 annually, the market price of the raw tobacco leaf, and the land, the labor and the capital expended in getting the product to the men who puff, breathe and perspire the noxious product into the air everyone must breathe, and who bespatter the streets, sidewalks, the floor of every public place and conveyance, and befoul the million spittoons, smoking rooms and smoking cars, all unnecessary and should be uncalled for, but whose installation and up-keep the non-user as well as the user is forced to pay, and this in a country of, for and by the people. This costly, filthy, selfish tobacco habit should be outgrown. Let it begin in every new home, where the mother helps the father in refusing to set the example, and let its indulgence be absolutely prohibited to everyone while in public school and to all in educational institutions.

Mr. League had been given a letter of introduction to one of the leading farmers of the village and it chanced that as we reached the entranceway to his home we were met by his son, just returning from the fields with his drill on his shoulder, and it is he standing in the illustration, Fig. 131, holding the letter of introduction in his hand. After we had taken this photograph and another one looking down the narrow street from the same point, we were led to the small open court of the home, perhaps forty by eighty feet, upon which all doors of the one-storied structures opened. It was dry and bare of everything green, but a row of very tall handsome trees, close relatives of our cottonwood, with trunks thirty feet to the limbs, looked down into the court over the roofs of the low thatched houses. Here we met the father and grand-father of the man with the drill, so that, with the boy carrying the baby in his arms, who had met his father in the street gateway, there were four generations of males at our conference. There were women and girls in the household but custom requires them to remain in retirement on such occasions.

A low narrow four-legged bench, not unlike our carpenter's saw-horse, five feet long, was brought into the court as a seat, which our host and we occupied in common. We had been similarly received at the home of Mrs. Wu in Chekiang province. On our right was the open doorway to the kitchen in which stood, erect and straight, the tall spare figure of the patriarch of the household, his eyes still shining black but with hair and long thin straggling beard a uniform dull ashen gray. No Chinese hair, it seems, ever becomes white with age. He seemed to have assumed the duties of cook for while we were there he lighted the fire in the kitchen and was busy, but was always the final oracle on any matter of difference of opinion between the younger men regarding answers to questions. Two sleeping apartments adjoining the kitchen, through whose wide kang beds the waste heat from the cooking was conveyed, as described on page 142, completed this side of the court. On our left was the main street completely shut off by a solid earth wall as high as the eaves of the house, while in front of us, adjoining the street, was the manure midden, a compost pit six feet deep and some eight feet square. A low opening in the street wall permitted the pit to be emptied and to receive earth and stubble or refuse from the fields for composting.

Fig. 135.—Farm village street with stacks of earth and piles of compost for use as fertilizer, Shantung, China.

Against the pit and without partition, but cut off from the court, was the home of the pigs, both under a common roof continuous with a closed structure joining with the sleeping apartments, while behind us and along the alley-way by which we had entered were other dwelling and storage compartments. Thus was the large family of four generations provided with a peculiarly private open court where they could work and come out for sun and air, both, from our standards, too meagerly provided in the houses.

We had come to learn more of the methods of fertilizing practiced by these people. The manure midden was before us and the piles of earth brought in from the fields, for use in the process, were stacked in the street, where we had photographed them at the entrance, as seen in Fig. 135. There a father, with his pipe, and two boys stand at the extreme left; beyond them is a large pile of earth brought into the village and carefully stacked in the narrow street; on the other side of the street, at the corner of the first building, is a pile of partly fermented compost thrown from a pit behind the walls. Further along in the street, on the same side, is a second large stack of soil where two boys are standing at either end and another little boy was in a near-by doorway. In front of the tree, on the left side of the street, stands a third boy, near him a small donkey and still another boy. Beyond this boy stands a third large stack of soil, while still beyond and across the way is another pile partly composted. Notwithstanding the cattle in the preceding illustration, the donkey, the men, the boys, the three long high stacks of soil and the two piles of compost, the ten rods of narrow street possessed a width of available travelway and a cleanliness which would appear impossible. Each farmer's household had its stack of soil in the street, and in walking through the village we passed dozens of men turning and mixing the soil and compost, preparing it for the field.

The compost pit in front of where we sat was two-thirds

filled. In it had been placed all of the manure and waste of the household and street, all stubble and waste roughage from the field, all ashes not to be applied directly and some of the soil stacked in the street. Sufficient water was added at intervals to keep the contents completely saturated and nearly submerged, the object being to control the character of fermentation taking place.

The capacity of these compost pits is determined by the amount of land served, and the period of composting is made as long as possible, the aim being to have the fiber of all organic material completely broken down, the result being a product of the consistency of mortar.

When it is near the time for applying the compost to the field, or of feeding it to the crop, the fermented product is removed in waterproof carrying baskets to the floor of the court, to the yard, such as seen in Fig. 126, or to the street, where it is spread to dry, to be mixed with fresh soil, more ashes, and repeatedly turned and stirred to bring about complete aeration and to hasten the processes of nitrification. During all of these treatments, whether in the compost pit or on the nitrification floor, the fermenting organic matter in contact with the soil is converting plant food elements into soluble plant food substances in the form of potassium, calcium and magnesium nitrates and soluble phosphates of one or another form, perhaps of the same bases and possibly others of organic type. If there is time and favorable temperature and moisture conditions for these fermentations to take place in the soil of the field before the crop will need it, the compost may be carried direct from the pit to the field and spread broadcast, to be plowed under. Otherwise the material is worked and reworked, with more water added if necessary, until it becomes a rich complete fertilizer, allowed to become dry and then finely pulverized, sometimes using stone rollers drawn over it by cattle, the donkey or by hand. The large numbers of stacks of compost seen in the fields between Tsingtao and Tsinan were of this type and thus laboriously prepared in the villages and then transported to the fields, stacked and plastered, to be ready for use at next planting.

In the early days of European history, before modern chemistry had provided the cheaper and more expeditious method of producing potassium nitrate for the manufacture of gunpowder and fireworks, much land and effort were devoted to niter-farming which was no other than a specific application of this most ancient Chinese practice and probably imported from China. While it was not until 1877 to 1879 that men of science came to know that the processes of nitrification, so indispensable to agriculture, are due to germ life, in simple justice to the plain farmers of the world, to those who through all the ages from Adam down, living close to Nature and working through her and with her, have fed the world, it should be recognized that there have been those among them who have grasped such essential, vital truths and have kept them alive in the practices of their day. And so we find it recorded in history as far back as 1686 that Judge Samuel Lewell copied upon the cover of his journal a practical man's recipé for making saltpeter beds, in which it was directed, among other things, that there should be added to it "mother of petre", meaning, in Judge Lewell's understanding, simply soil from an old niter bed, but in the mind of the man who applied the maternity prefix, —mother,—it must have meant a vital germ contained in the soil, carried with it, capable of reproducing its kind and of perpetuating its characteristic work, belonging to the same category with the old, familiar, homely germ, "mother" of vinegar. So, too, with the old cheesemaker who grasped the conception which led to the long time practice of washing the walls of a new cheese factory with water from an old factory of the same type, he must have been led by analogies of experience with things seen to realize that he was here dealing with a vital factor. Hundreds, of course, have practiced empyrically, but some one preceded with the essential thought and we feel it is small credit to men of our time who, after ten or twenty

years of technical training, having their attention directed
to a something to be seen, and armed with compound micro-
scopes which permit them to see with the physical eye
the "mother of petre", arrogate to themselves the discovery of a great truth. Much more modest would it be and
much more in the spirit of giving credit where credit is
due to admit that, after long doubting the existence of
such an entity, we have succeeded in confirming in fullness
the truth of a great discovery which belongs to an unnamed
genius of the past, or perhaps to a hundred of them who,
working with life's processes and familiar with them
through long intimate association, saw in these invisible
processes analogies that revealed to them the essential
truth in such fullness as to enable them to build upon it
an unfailing practice.

There is another practice followed by the Chinese, connected with the formation of nitrates in soils, which again
emphasizes the national trait of saving and turning to use
any and every thing worth while. Our attention was
called to this practice by Rev. A. E. Evans of Shunking,
Szechwan province. It rests upon the tendency of the
earth floors of dwellings to become heavily charged with
calcium nitrate through the natural processes of nitrification. Calcium nitrate being deliquescent absorbs moisture
sufficiently to dissolve and make the floor wet and sticky.
Dr. Evans' attention was drawn to the wet floor in his
own house, which he at first ascribed to insufficient ventilation, but which he was unable to remedy by improving
that. The father of one of his assistants, whose business
consisted in purchasing the soil of such floors for producing potassium nitrate, used so much in China in the manufacture of fireworks and gunpowder, explained his difficulty and suggested the remedy.

This man goes from house to house through the village,
purchasing the soil of floors which have thus become over-
charged. He procures a sample, tests it and announces
what he will pay for the surface two, three or four inches,
the price sometimes being as high as fifty cents for the

privilege of removing the top layer of the floor, which the proprietors must replace. He leaches the soil removed, to recover the calcium nitrate, and then pours the leachings through plant ashes containing potassium carbonate, for the purpose of transforming the calcium nitrate into the potassium nitrate or saltpeter. Dr. Evans learned that during the four months preceding our interview this man had produced sufficient potassium nitrate to bring his sales up to $80, Mexican. It was necessary for him to make a two-days journey to market his product. In addition he paid a license fee of 80 cents per month. He must purchase his fuel ashes and hire the services of two men.

When the nitrates which accumulate in the floors of dwellings are not collected for this purpose the soil goes to the fields to be used directly as a fertilizer, or it may be worked into compost. In the course of time the earth used in the village walls and even in the construction of the houses may disintegrate so as to require removal, but in all such cases, as with the earth brick used in the kangs. the value of the soil has improved for composting and is generally so used. This improvement of the soil will not appear strange when it is stated that such materials are usually from the subsoil, whose physical condition would improve when exposed to the weather, converting it in fact into an uncropped virgin soil.

We were unable to secure definite data as to the chemical composition of these composts and cannot say what amounts of available plant food the Shantung farmers are annually returning to their fields. There can be little doubt, however, that the amounts are quite equal to those removed by the crops. The soils appeared well supplied with organic matter and the color of the foliage and the general aspect of crops indicated good feeding.

The family with whom we talked in the village place their usual yields of wheat at 420 catty of grain and 1000 catty of straw per mow,* the grain being worth 35 strings

* Their mow was four-thirds of the legal standard mow.

of cash and the straw 12 to 14 strings, a string of cash being 40 cents, Mexican, at this time. Their yields of beans were such as to give them a return of 30 strings of cash for the grain and 8 to 10 strings for the straw. Small millet usually yielded 450 catty of grain, worth 25 strings of cash, per mow, and 800 catty of straw worth 10 to 11 strings of cash; while the yields of large millet they placed at 400 catty per mow, worth 25 strings of cash, and 1000 catty of straw worth 12 to 14 strings of cash. Stating these amounts in bushels per acre and in our currency, the yield of wheat was 42 bushels of grain and 6000 pounds of straw per acre, having a cash value of $27.09 for the grain and $10.06 for the straw. The soy bean crop follows the wheat, giving an additional return of $23.22 for the beans and $6.97 for the straw, making the gross earning for the two crops $67.34 per acre. The yield of small millet was 54 bushels of seed and 4800 pounds of straw per acre, worth $27.09 and $8.12 for seed and straw respectively, while the kaoliang or large millet gave a yield of 48 bushels of grain and 6000 pounds of stalks per acre, worth $19.35 for the grain, and $10.06 for the straw.

A crop of wheat like the one stated, if no part of the plant food contained in the grain or straw were returned to the field, would deplete the soil to the extent of about 90 pounds of nitrogen, 15 pounds of phosphorus and 65 pounds of potassium; and the crop of soy beans, if it also were entirely removed, would reduce these three plant food elements in the soil to the extent of about 240 pounds of nitrogen, 33 pounds of phosphorus and 102 pounds of potassium, on the basis of 45 bushels of beans and 5400 pounds of stems and leaves per acre, assuming that the beans added no nitrogen to the soil, which is of course not true. This household of farmers, therefore, in order to have maintained this producing power in their soil, have been compelled to return to it annually, in one form or another, not less than 48 pounds of phosphorus and 167 pounds of potassium per acre. The 330 pounds of nitrogen they would have to return in the form of organic matter or accumulate it from the atmosphere, through the instrumentality of their soy bean crop or some other legume. It has already been stated that they do add more than 5000 to 7000 pounds of dry compost, which, repeated for a second crop, would make an annual application of five to seven tons of dry compost per acre annually. They do use, in addition to this compost, large amounts of bean and peanut cake, which carry all of the plant food elements derived from the soil which are contained in the beans and the

Fig. 136.—Stone mill for grinding soy beans and peanuts, Shantung, China.

peanuts. If the vines are fed, or if the stems of the beans are burned for fuel, most of the plant food elements in these will be returned to the field, and they have doubtless learned how to completely restore the plant food elements removed by their crops, and persistently do so.

The roads made by the Germans in the vicinity of Tsingtao enabled us to travel by ricksha into the adjoining country, and on one such trip we visited a village mill for grinding soy beans and peanuts in the manufacture of oil, and Fig. 136 shows the stone roller, four feet in diameter and two feet thick, which is revolved about a vertical axis on a circular stone plate, drawn by a donkey, crushing the kernels partly by its weight and partly by a twisting motion, for the arm upon which the roller revolves is very short. After the meal had been ground the oil was expressed in essentially the same way as that described for the cotton seed, but the bean and peanut cakes are made much larger than the cotton seed cakes, about eighteen inches in diameter and three to four inches thick. Two of these cakes are seen in Fig. 137, standing on edge outside

Fig. 137.—Two large peanut cakes and a paper demijohn for containing the oil, outside the village mill, Shantung, China.

the mill in an orderly clean court. It is in this form that bean cake is exported in large quantities to different parts of China, and to Japan in recent years, for use as fertilizer, and very recently it is being shipped to Europe for both stock food and fertilizer.

Nowhere in this province, nor further north, did we see the large terra cotta receptacles so extensively used in the south for storing human excreta. In these dryer climates some method of desiccation is practiced and we found the gardeners in the vicinity of Tsingtao with quantities of the fertilizer stacked under matting shelters in

17

the desiccated condition, this being finely pulverized in one or another way before it was applied. The next illustration, Fig. 138, shows one of these piles being fitted for the garden, its thatched shelter standing behind the grandfather of a household. His grandson was carrying the prepared fertilizer to the garden area seen in Fig. 139, where the father was working it into the soil. The greatest pains is taken, both in reducing the product to a fine powder and in spreading and incorporating it with the soil, for one of their maxims of soil management is to

Fig. 138.—Pulverizing desiccated human excreta preparatory for use in garden fertilization, Shantung, China.

make each square foot of field or garden the equal of every other in its power to produce. In this manner each little holding is made to yield the highest returns possible under the conditions the husbandman is able to control.

From one portion of the area being fitted, a crop of artemisia had been harvested, giving a gross return at the rate of $73.19 per acre, and from another leeks had been taken, bringing a gross return of $43.86 per acre. Chinese celery was the crop for which the ground was being fitted.

The application of soil as a fertilizer to the fields of

China, whether derived from the subsoil or from the silts
and organic matter of canals and rivers, must have played
an important part in the permanency of agriculture in
the Far East, for all such additions have been positive
accretions to the effective soil, increasing its depth and
carrying to it all plant food elements. If not more than
one-half of the weight of compost applied to the fields

Fig. 139.—Gardener thoroughly incorporating fertilizer with his soil preparatory to planting a second crop of the season, May 24th, Shantung, China.

of Shantung is highly fertilized soil, the rates of application observed would, in a thousand years, add more than
two million pounds per acre, and this represents about
the volume of soil we turn with the plow in our ordinary
tillage operations, and this amount of good soil may carry
more than 6000 pounds of nitrogen, 2000 pounds of phosphorus and more than 60,000 pounds of potassium.

When we left our hotel by ricksha for the steamer, returning to Shanghai, we soon observed a boy of thirteen or fourteen years apparently following, sometimes a little ahead, sometimes behind, usually keeping the sidewalk but slackening his pace whenever the ricksha man came to a walk. It was a full mile to the wharf. The boy evidently knew the sailing schedule and judged by the valise in front, that we were to take the out-going steamer and that he might possibly earn two cents, Mexican, the usual fee for taking a valise aboard the steamer. Twenty men at the wharf might be waiting for the job, but he was taking the chance with the mile down and back thrown in, and all for less than one cent in our currency, equivalent at the time to about twenty "cash". As we neared the steamer the lad closed up behind but strong and eager men were watching. Twice he was roughly thrust aside and before the ricksha stopped a man of stalwart frame seized the valise and, had we not observed the boy thus unobtrusively entering the competition, he would have had only his trouble for his pains. Thus intense was the struggle here for existence and thus did a mere lad put himself effectively into it. True to breeding and example he had spared no labor to win and was surprised but grateful to receive more than he had expected.

XI.

ORIENTALS CROWD BOTH TIME AND SPACE.

Time is a function of every life process, as it is of every physical, chemical and mental reaction, and the husbandman is compelled to shape his operations so as to conform with the time requirements of his crops. The oriental farmer is a time economizer beyond any other. He utilizes the first and last minute and all that are between. The foreigner accuses the Chinaman of being always ''long on time'', never in a fret, never in a hurry. And why should he be when he leads time by the forelock, and uses all there is?

The customs and practices of these Farthest East people regarding their manufacture of fertilizers in the form of earth composts for their fields, and their use of altered subsoils which have served in their kangs, village walls and dwellings, are all instances where they profoundly shorten the time required in the field to affect the necessary chemical, physical and biological reactions which produce from them plant food substances. Not only do they thus increase their time assets, but they add, in effect, to their land area by producing these changes outside their fields, at the same time giving their crops the immediately active soil products.

Their compost practices have been of the greatest consequence to them, both in their extremely wet, rice-culture methods, and in their ''dry-farming'' practices, where the soil moisture is too scanty during long periods to permit rapid fermentation under field conditions. Western agriculturalists have not sufficiently appreciated the fact that the most rapid growth of plant food substances in the soil cannot occur at the same time and place with the most rapid crop increase, because both processes draw upon the available soil moisture, soil air and soluble potassium, calcium, phosphorus and nitrogen compounds. Whether this fundamental principle of practical agriculture is written in their literature or not it is most indelibly fixed

Fig. 140.—Looking across reservoir and four-man foot-power pump, used to lift water to a nursery rice bed, at fields of grain sowed broadcast in narrow beds.

in their practice. If we and they can perpetuate the essentials of this practice at a large saving of human effort, or perpetually secure the final result in some more expeditious and less laborious way, most important progress will have been made.

When we went north to the Shantung province the Kiangsu and Chekiang farmers were engaged in another of their time saving practices, also involving a large amount of human labor. This was the planting of cotton in wheat fields before the wheat was quite ready to harvest. In the sections of these two provinces which we visited most of the wheat and barley were sowed broadcast on narrow raised lands, some five feet wide, with furrows between, after the manner seen in Fig. 140, showing a reservoir in the immediate foreground, on whose bank is installed one of the four-man foot-power irrigation pumps in use to flood the nursery rice bed close by on the right. The narrow lands of broadcasted wheat extend back from

Fig. 141.—Field of wheat with grain four feet, eight inches high, nearing time of harvest, in which cotton is planted.

the reservoir toward the farmsteads which dot the landscape, and on the left stands one of the pump shelters near the canal bank.

To save time, or lengthen the growing season of the cotton which was to follow, this seed was sown broadcast among the grain on the surface, some ten to fifteen days before the wheat would be harvested. To cover the seed the soil in the furrows between the beds had been spaded loose to a depth of four or five inches, finely pulverized, and then with a spade was evenly scattered over the bed, letting it sift down among the grain, covering the seed. This loose earth, so applied, acts as a mulch to conserve the capillary moisture, permitting the soil to become sufficiently damp to germinate the seed before the wheat is harvested. The next illustration, Fig. 141, is a closer view with our interpreter standing in another field of wheat in which cotton was being sowed April 22nd in the manner described, and yet the stand of grain was

Fig. 142.—View of same field as Fig. 141, after the grain had been cut, removed and the cotton sowed in it was up.

very close and shoulder high, making it not an easy task either to sow the seed or to scatter sufficient soil to cover it.

When we had returned from Shantung this piece of grain had been harvested, giving a yield of 95.6 bushels of wheat and 3.5 tons of straw per acre, computed from the statement of the owner that 400 catty of grain and 500 catty of straw had been taken from the beds measuring 4050 square feet. On the morning of May 29th the photograph for Fig. 142 was taken, showing the same area after the wheat had been harvested and the cotton was up, the young plants showing slightly through the short stubble. These beds had already been once treated with liquid fertilizer. A little later the plants would be hoed and thinned to a stand of about one plant per each

Fig. 143.—Multiple crops, wheat, windsor beans and cotton. Wheat ready to harvest, beans two-thirds grown, cotton just planted. Upper view looking between wheat rows, lower, looking between bean rows now covering ground.

square foot of surface. There were thirty-seven days between the taking of the two photographs, and certainly thirty days had been added to the cotton crop by this method of planting, over what would have been available if the grain had been first harvested and the field fitted before planting. It will be observed that the cotton follows the wheat without plowing, but the soil was deep, naturally open, and a layer of nearly two inches of loose earth had been placed over the seed at the time of planting. Besides, the ground would be deeply worked with the two or four tined hoe, at the time of thinning.

Starting cotton in the wheat in the manner described is but a special case of a general practice widely in vogue. The growing of multiple crops is the rule throughout these countries wherever the climate permits. Sometimes as many as three crops occupy the same field in recurrent

Fig. 144.—Turning under a crop of "Chinese clover" for green manure, grown with barley and to be followed by cotton.

rows, but of different dates of planting and in different stages of maturity. Reference has been made to the overlapping and alternation of cucumbers with greens. The general practice of planting nearly all crops in rows lends itself readily to systems of multiple cropping, and these to the fullest possible utilization of every minute of the growing season and of the time of the family in caring for the crops. In the field, Fig. 143, a crop of winter wheat was nearing maturity, a crop of windsor beans was about two-thirds grown, and cotton had just been planted, April 22nd. This field had been thrown into ridges some five feet wide with a twelve inch furrow between them. Two rows of wheat eight inches wide, planted two feet between centers cccupied the crest of the ridge, leaving a strip sixteen inches wide, seen in the upper section, (1) for tillage, (2) then fertilization and (3) finally the row of cotton planted just before the wheat was harvested. Against the furrow on each side was a row of windsor beans, seen in the lower view, hiding the furrow, which was matured some time after the wheat was harvested and before the cotton was very large. A late fall crop sometimes follows the windsor beans after a period of tillage

Fig. 145.—Multiple crops in Chihli—wheat and sorghum, the wheat ripe, to be followed by soy beans. Piles of compost earth for soy beans.

and fertilization, making four in one year. With such a succession fertilization for each crop, and an abundance of soil moisture are required to give the largest returns from the soil.

In another plan winter wheat or barley may grow side by side with a green crop, such as the 'Chinese clover'' (*Medicago denticulata*, Willd.) for soil fertilizer, as was the case in Fig. 144, to be turned under and fertilize for a crop of cotton planted in rows on either side of a crop of barley. After the barley had been harvested the ground it occupied would be tilled and further fertilized, and when the cotton was nearing maturity a crop of rape might be grown, from which "salted cabbage" would be prepared for winter use.

Multiple crops are grown as far north in Chihli as Tientsin and Peking, these being oftenest wheat, maize, large and small millet and soy beans, and this, too, where the soil is less fertile and where the annual rainfall is only about twenty-five inches, the rainy season beginning in late June or early July, and Fig. 145 shows one of

Fig. 146.—Family engaged in cutting, from bundles of wheat, the roots to be used in making compost, Chihli, China.

these fields as it appeared June 14th, where two rows of wheat and two of large millet were planted in alternating pairs, the rows being about twenty-eight inches apart. The wheat was ready to harvest but the straw was unusually short because growing on a light sandy loam in a season of exceptional drought, but little more than two inches of rain having fallen after January 1st of that year.

The piles of pulverized dry-earth compost seen between the rows had been brought for use on the ground occupied by the wheat when that was removed. The wheat would be pulled, tied in bundles, taken to the village and the roots cut off, for making compost, as in Fig. 146, which shows the family engaged in cutting the roots from the small bundles of wheat, using a long straight knife blade, fixed at one end, and thrust downward upon the bundle with lever pressure. These roots, if not used as fuel, would be transferred to the compost pit in the enclosure seen in Fig. 147, whose walls were built of earth brick. Here, with any other waste litter, manure or ashes, they

Fig. 147.—Compost shelter and pig pen, with pile of wheat roots stacked at one end, for use in making compost, Chihli, China.

would be permitted to decay under water until the fiber had been destroyed, thus permitting it to be incorporated with soil and applied to the fields, rich in soluble plant food and in a condition which would not interfere with the capillary movement of soil moisture, the work going on outside the field where the changes could occur unimpeded and without interfering with the growth of crops on the ground.

In this system of combined intertillage and multiple cropping the oriental farmer thus takes advantage of whatever good may result from rotation or succession of crops, whether these be physical, vito-chemical or biological. If plants are mutually helpful through close association of their root systems in the soil, as some believe may be the case, this growing of different species in close juxtaposition would seem to provide the opportunity, but the other advantages which have been pointed out are so evident and so important that they, rather than this, have doubtless led to the practice of growing different crops in close recurrent rows.

XII.

RICE CULTURE IN THE ORIENT.

The basal food crop of the people of China, Korea and Japan is rice, and the mean consumption in Japan, for the five years ending 1906, per capita and per annum, was 302 pounds. Of Japan's 175,428 square miles she devoted, in 1906, 12,856 to the rice crop. Her average yield of water rice on 12,534 square miles exceeded 33 bushels per acre, and the dry land rice averaged 18 bushels per acre on 321 square miles. In the Hokkaido, as far north as northern Illinois, Japan harvested 1,780,000 bushels of water rice from 53,000 acres.

In Szechwan province, China, Consul-General Hosie places the yield of water rice on the plains land at 44 bushels per acre, and that of the dry land rice at 22 bushels. Data given us in China show an average yield of 42 bushels of water rice per acre, while the average yield of wheat was 25 bushels per acre, the normal yield in Japan being about 17 bushels.

If the rice eaten per capita in China proper and Korea is equal to that in Japan the annual consumption for the three nations, using the round number 300 pounds per capita per annum, would be:

	Population.	Consumption.
China	410,000,000	61,500,000 tons
Korea	12,000,000	1,800,000 tons
Japan	53,000,000	7,950 000 tons
Total	475,000,000	71,250,000 tons

If the ratio of irrigated to dry land rice in Korea and China proper is the same as that in Japan, and if the

mean yield of rice per acre in these countries were forty
bushels for the water rice and twenty bushels for the dry
land rice, the acreage required to give this production
would be:

	Area.	
	Water rice, sq. miles.	Dry land rice, sq. miles.
In China	78,073	4,004
In Korea	2,285	117
In Japan	12,534	321
Sum	92,892	4,442
Total	97,334	

Our observations along the four hundred miles of railway in Korea between Antung, Seoul and Fusan, suggest
that the land under rice in this country must be more
rather than less than that computed, and the square miles
of canalized land in China, as indicated on pages 97 to
102, would indicate an acreage of rice for her quite as large
as estimated.

In the three main islands of Japan more than fifty per
cent of the cultivated land produces a crop of water rice
each year and 7.96 per cent of the entire land area of the
Empire, omitting far-north Karafuto. In Formosa and in
southern China large areas produce two crops each year.
At the large mean yield used in the computation the estimated acreage of rice in China proper amounts to 5.93
per cent of her total area and this is 7433 square miles
greater than the acreage of wheat in the United States
in 1907. Our yield of wheat, however, was but
19,000,000 tons, while China's output of rice was certainly
double and probably three times this amount from nearly
the same acreage of land; and notwithstanding this large
production per acre, more than fifty per cent, possibly
as high as seventy-five per cent, of the same land matures
at least one other crop the same year, and much of this
may be wheat or barley, both chiefly consumed as human
food.

Had the Mongolian races spread to and developed in
North America instead of, or as well as, in eastern Asia,

there might have been a Grand Canal, something as suggested in Fig. 148, from the Rio Grande to the mouth of the Ohio river and from the Mississippi to Chesapeake Bay, constituting more than two thousand miles of inland water-way, serving commerce, holding up and redistributing both the run-off water and the wasting fertility of soil erosion, spreading them over 200,000 square miles of thoroughly canalized coastal plains, so many of which are

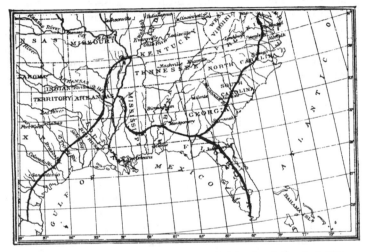

Fig. 148.—A canal which would correspond with the Grand Canal of China.

now impoverished lands, made so by the intolerable waste of a vaunted civilization. And who shall venture to enumerate the increase in the tonnage of sugar, bales of cotton, sacks of rice, boxes of oranges, baskets of peaches, and in the trainloads of cabbage, tomatoes and celery such husbanding would make possible through all time; or number the increased millions these could feed and clothe? We may prohibit the exportation of our phosphorus, grind our limestone, and apply them to our fields, but this alone is only temporizing with the future. The more we produce, the more numerous our millions, the faster must present practices speed the waste to the sea, from whence neither money nor prayer can call them back.

If the United States is to endure; if we shall project our history even through four or five thousand years as the Mongolian nations have done, and if that history shall be written in continuous peace, free from periods of wide spread famine or pestilence, this nation must orient itself; it must square its practices with a conservation of resources which can make endurance possible. Intensifying cultural methods but intensifies the digestion, assimilation and exhaustion of the surface soil, from which life springs. Multiple cropping, closer stands on the ground and stronger growth, all mean the transpiration of much more water per acre through the crops, and this can only be rendered possible through a redistribution of the run-off and the adoption of irrigation practices in humid climates where water exists in abundance. Sooner or later we must adopt a national policy which shall more completely conserve our water resources, utilizing them not only for power and transportation, but primarily for the maintenance of soil fertility and greater crop production through supplemental irrigation, and all these great national interests should be considered collectively, broadly, and with a view to the fullest and best possible coordination. China, Korea and Japan long ago struck the keynote of permanent agriculture but the time has now come when they can and will make great improvements, and it remains for us and other nations to profit by their experience, to adopt and adapt what is good in their practice and help in a world movement for the introduction of new and improved methods.

In selecting rice as their staple crop; in developing and maintaining their systems of combined irrigation and drainage, notwithstanding they have a large summer rainfall; in their systems of multiple cropping; in their extensive and persistent use of legumes; in their rotations for green manure to maintain the humus of their soils and

Fig. 149.—Recently transplanted rice fields in Japan.

for composting; and in the almost religious fidelity with which they have returned to their fields every form of waste which can replace plant food removed by the crops, these nations have demonstrated a grasp of essentials and of fundamental principles which may well cause western nations to pause and reflect.

While this country need not and could not now adopt their laborious methods of rice culture, and while, let us

Fig. 150.—Rice fields on the plains of the Yangtse kiang, China, being flooded preparatory to transplanting rice.

hope, those who come after us may never be compelled to do so, it is nevertheless quite worth while to study, for the sake of the principles involved, the practices they have been led to adopt.

Great as is the acreage of land in rice in these countries, but little, relatively, is of the dry land type, and the fields upon which most of the rice grows have all been graded to a water level and surrounded by low, narrow raised rims, such as may be seen in Fig. 149 and in Fig. 150, where three men are at work on their foot-power pump, flooding fields preparatory to transplanting the rice. If the country was not level then the slopes have been graded into horizontal terraces varying in size according to the steepness of the areas in which they were cut. We saw these often no larger than the floor of a small room, and Professor Ross informed me that he walked past those in the interior of China no larger than a dining table and that he saw one bearing its crop of rice, surrounded by its rim and holding water, yet barely larger than a good napkin. The average area of the paddy field in Japan is officially reported at 1.14 se, or an area of but 31 by 40 feet. Excluding Hokkaido, Formosa and Karafuto, fifty-three per cent of the irrigated rice lands in Japan are in allotments smaller than one-eighth of an acre, and seventy-four per cent of other cultivated lands are held in areas less than one-fourth of an acre, and each of these may be further subdivided. The next two illustrations, Figs. 151 and 152, give a good idea both of the small size of the rice fields and of the terracing which has been done to secure the water level basins. The house standing near the center of Fig. 151 is a good scale for judging both the size of the paddies and the slope of the valley. The distance between the rows of rice is scarcely one foot, hence counting these in the foreground may serve as another measure. There are more than twenty little fields shown in this engraving in front of the house and reaching but half way to it, and the house was less than five hundred feet from the camera.

There are more than eleven thousand square miles of fields thus graded in the three main islands of Japan, each provided with rims, with water supply and drainage channels, all carefully kept in the best of repair. The more level areas, too, in each of the three countries, have been similarly thrown into water level basins, comparatively few of which cover large areas, because nearly always the holdings are small. All of the earth excavated from the canals and drainage channels has been leveled over the fields unless needed for levees or dikes, so that

Fig. 151.—Terraced valley with small terraces flooded and transplanted to rice, Japan.

the original labor of construction, added to that of maintenance, makes a total far beyond our comprehension and nearly all of it is the product of human effort.

The laying out and shaping of so many fields into these level basins brings to the three nations an enormous aggregate annual asset, a large proportion of which western nations are not yet utilizing. The greatest gain

Fig. 152.—Looking down a steep, narrow Japanese valley at small, flooded and transplanted rice paddies.

comes from the unfailing higher yields made possible by providing an abundance of water through which more plant food can be utilized, thus providing higher average yields. The waters used, coming as they do largely from the uncultivated hills and mountain lands, carrying both dissolved and suspended matters, make positive annual additions of dissolved limestone and plant food elements to the fields which in the aggregate have been very

large, through the persistent repetitions which have prevailed for centuries. If the yearly application of such
water to the rice fields is but sixteen inches, and this has
the average composition quoted by Merrill for rivers of
North America, taking into account neither suspended
matter nor the absorption of potassium and phosphorus
by it, each ten thousand square miles would receive, dissolved in the water, substances containing some 1,400
tons of phosphorus; 23,000 tons of potassium; 27,000
tons of nitrogen; and 48,000 tons of sulphur. In addition, there are brought to the fields some 216,000 tons
of dissolved organic matter and a still larger weight of
dissolved limestone, so necessary in neutralizing the acidity of soils, amounting to 1,221,000 tons; and such
savings have been maintained in China, Korea and Japan
on more than five, and possibly more than nine, times
the ten thousand square miles, through centuries. The
phosphorus thus turned upon ninety thousand square
miles would aggregate nearly thirteen million tons in a
thousand years, which is less than the time the practice
has been maintained, and is more phosphorus than would
be carried in the entire rock phosphate thus far mined
in the United States, were it all seventy-five per cent
pure.

The canalization of fifty thousand square miles of our
Gulf and Atlantic coastal plain, and the utilization on
the fields of the silts and organic matter, together with
the water, would mean turning to account a vast tonnage
of plant food which is now wasting into the sea, and a
correspondingly great increase of crop yield. There
ought, and it would seem there must some time be provided a way for sending to the sandy plains of Florida,
and to the sandy lands between there and the Mississippi,
large volumes of the rich silt and organic matter from this
and other rivers, aside from that which should be applied
systematically to building above flood plain the lands of
the delta which are subject to overflow or are too low to
permit adequate drainage.

It may appear to some that the application of such large volumes of water to fields, especially in countries of heavy rainfall, must result in great loss of plant food through leaching and surface drainage. But under the remarkable practices of these three nations this is certainly not the case and it is highly important that our people should understand and appreciate the principles which underlie the practices they have almost uniformly adopted on

Fig. 153 —Egg plants growing in the midst of rice fields with soil continually saturated and water standing in surface drain within 14 inches of the surface, Japan.

the areas devoted to rice irrigation. In the first place, their paddy fields are under-drained so that most of the water either leaves the soil through the crop, by surface evaporation, or it percolates through the subsoil into shallow drains. When water is passed directly from one rice paddy to another it is usually permitted some time after fertilization, when both soil and crop have had time to appropriate or fix the soluble plant food substances. Besides this, water is not turned upon the fields until the

time for transplanting the rice, when the plants are already
provided with a strong root system and are capable of at
once appropriating any soluble plant food which may
develop about their roots or be carried downward over
them.

Although the drains are of the surface type and but
eighteen inches to three feet in depth, they are sufficiently
numerous and close so that, although the soil is continuously nearly filled with water, there is a steady percola-

Fig. 154.—Watermelons, with the ground heavily mulched with straw, growing
on low beds under conditions similar to those of Fig. 153.

tion of the fresh, fully aerated water carrying an abundance of oxygen into the soil to meet the needs of the
roots, so that watermelons, egg plants, musk melons and
taro are grown in the rotations on the small paddies
among the irrigated rice after the manner seen in the illustrations. In Fig. 153 each double row of egg plants is
separated from the next by a narrow shallow trench
which connects with a head drain and in which water was
standing within fourteen inches of the surface. The
same was true in the case of the watermelons seen in
Fig. 154, where the vines are growing on a thick layer of

straw mulch which holds them from the moist soil and acts to conserve water by diminishing evaporation and, through decay from the summer rains and leaching, serves as fertilizer for the crop. In Fig. 155 the view is along a pathway separating two head ditches between areas in watermelons and taro, carrying the drainage waters from the several furrows into the main ditches. Although the soil appeared wet the plants were vigorous and

Fig. 155.—Looking along a path between two head ditches separating patches of watermelons and taro, Japan.

healthy, seeming in no way to suffer from insufficient drainage.

These people have, therefore, given effective attention to the matter of drainage as well as irrigation and are looking after possible losses of plant food, as well as ways of supplying it. It is not alone where rice is grown that cultural methods are made to conserve soluble plant food and to reduce its loss from the field, for very often, where flooding is not practiced, small fields and beds, made quite level, are surrounded by low raised borders which permit not only the whole of any rain to be retained upon the field when so desired, but it is completely distributed over it, thus causing the whole soil to be uniformly charged with moisture and preventing washing from one portion of the field to another. Such provisions are shown in Figs. 133 and 138.

Extensive as is the acreage of irrigated rice in China, Korea and Japan, nearly every spear is transplanted; the largest and best crop possible, rather than the least labor and trouble, as is so often the case with us, determining their methods and practices. We first saw the fitting of

Fig. 156.—Residence compound and farm buildings of Mrs. Wu, Kashing, China.

the rice nursery beds at Canton and again near Kashing in Chekiang province on the farm of Mrs. Wu, whose homestead is seen in Fig. 156. She had come with her husband from Ningpo after the ravages of the Taiping rebellion had swept from two provinces alone twenty millions of people and settled on a small area of then vacated land. As they prospered they added to their holding by purchase until about twenty-five acres were acquired, an area about ten times that possessed by the usual prosperous family in China. The widow was managing her place, one of her sons, although married, being still in school, the daughter-in-law living with her mother-in-law and helping in the home. Her field help during the summer consisted of seven laborers and she kept four cows for the plowing and pumping of water for irrigation. The wages of the men were at the rate of $24, Mexican, for five summer months, together with their meals which were four each day. The cash outlay for the seven men was thus $14.45 of our currency per month. Ten years before, such labor had been $30 per year, as compared with $50 at the time of our visit, or $12.90 and $21.50 of our currency, respectively.

Her usual yields of rice were two piculs per mow, or twenty-six and two-thirds bushels per acre, and a wheat crop

Fig. 157.—Pumping station on the farm of Mrs. Wu, showing pump shelter, two power wheels connected with pumps, set at the end of a water channel leading from a canal.

yielding half this amount, or some other, was taken from part of the land the same season, one fertilization answering for the two crops. She stated that her annual expense for fertilizers purchased was usually about $60, or $25.80 of our currency. The homestead of Mrs. Wu, Fig. 156, consists of a compound in the form of a large quadrangle surrounding a court closed on the south by a solid wall eight feet high. The structure is of earth brick with the roof thatched with rice straw.

Our first visit here was April 19th. The nursery rice beds had been planted four days, sowing seed at the rate of twenty bushels per acre. The soil had been very carefully prepared and highly fertilized, the last treatment being a dressing of plant ashes so incompletely burned as to leave the surface coal black. The seed, scattered directly upon the surface, almost completely covered it and had been gently beaten barely into the dressing of ashes, using a wide, flat-bottom basket for the purpose. Each evening, if the night was likely to be cool, water was pumped over the bed, to be withdrawn the next day,

Fig. 158.—Close view of power wheel with cow attached, used in driving the irrigation pump, one of the two seen in Fig. 157.

if warm and sunny, permitting the warmth to be absorbed by the black surface, and a fresh supply of air to be drawn into the soil.

Nearly a month later, May 14th, a second visit was made to this farm and one of the nursery beds of rice, as it then appeared, is seen in Fig. 159, the plants being about eight inches high and nearing the stage for transplanting. The field beyond the bed had already been partly flooded and plowed, turning under "Chinese clover" to ferment as green manure, preparatory for the rice transplanting. On the opposite side of the bed and

Fig. 159.—Nursery bed of rice 29 days planted, showing irrigation furrows; field beyond flooded, partly plowed, and the rice nearly ready for transplanting.

in front of the residence, Fig. 156, flooding was in progress in the furrows between the ridges formed after the previous crop of rice was harvested and upon which the crop of clover for green manure was grown. Immediately at one end of the two series of nursery beds, one of which is seen in Fig. 159, was the pumping plant seen in Fig. 157, under a thatched shelter, with its two pumps installed at the end of a water channel leading from the canal. One of these wooden pump powers, with the blind-

Fig. 160.—Plowed field nearly fitted for rice, and the smoothing, pulverizing harrow used for the purpose, Chekiang province, China.

folded cow attached, is reproduced in Fig. 158 and just beyond the animal's head may be seen the long handle dipper to which reference has been made, used for collecting excreta.

More than a month is saved for maturing and harvesting winter and early spring crops, or in fitting the fields for rice, by this planting in nursery beds. The irrigation period for most of the land is cut short a like amount, saving in both water and time. It is cheaper and easier to highly fertilize and prepare a small area for the nursery, while at the same time much stronger and more uniform plants are secured than would be possible by sowing in the field. The labor of weeding and caring for the plants in the nursery is far less than would be required in the field. It would be practically impossible to fit the entire rice areas as early in the season as the nursery beds are fitted, for the green manure is not yet grown and time is required for composting or for decaying, if plowed under directly. The rice plants in the nursery are carried to a stage when they are strong feeders and when set into the

Fig. 161.—Form of revolving wooden harrow for fitting flooded rice fields preparatory to transplanting.

newly prepared, fertilized, clean soil of the field they are ready to feed strongly under these most favorable conditions Both time and strength of plant are thus gained and these people are following what would appear to be the best possible practices under their condition of small holdings and dense population.

With our broad fields, our machinery and few people, their system appears to us crude and impossible, but cut our holdings to the size of theirs and the same stroke makes our machinery, even our plows, still more impossible, and so the more one studies the environment of these people, thus far unavoidable, their numbers, what they have done and are doing, against what odds they have succeeded, the more difficult it becomes to see what course might have been better.

How full with work is the month which precedes the transplanting of rice has been pointed out,—the making of the compost fertilizer; harvesting the wheat, rape and beans; distributing the compost over the fields, and their flooding and plowing. In Fig. 160 one of these fields is

Fig. 162.—Group of Chinese women pulling rice in a nursery bed, tying the plants in bundles preparatory to transplanting.

seen plowed, smoothed and nearly ready for the plants. The turned soil had been thoroughly pulverized, leveled and worked to the consistency of mortar, on the larger fields with one or another sort of harrow, as seen in Figs. 160 and 161. This thorough puddling of the soil permits the plants to be quickly set and provides conditions which ensure immediate perfect contact for the roots.

When the fields are ready women repair to the nurseries with their low four-legged bamboo stools, to pull the rice plants, carefully rinsing the soil from the roots, and then tie them into bundles of a size easily handled in transplanting, which are then distributed in the fields.

Fig. 163.—Transplanting rice in China. Four views taken from the same point at intervals of fifteen minutes, showing the progress made during forty-five minutes.

The work of transplanting may be done by groups of families changing work, a considerable number of them laboring together after the manner seen in Fig. 163, made from four snap shots taken from the same point at intervals of fifteen minutes. Long cords were stretched in the rice field six feet apart and each of the seven men was setting six rows of rice one foot apart, six to eight plants in a hill, and the hills eight or nine inches apart in the row. The bundle was held in one hand and deftly, with the other, the desired number of plants were selected with the fingers at the roots, separated from the rest and, with a single thrust, set in place in the row. There was no packing of earth about the roots, each hill being set with a single motion, which followed one another in quick succession, completing one cross row of six hills after another. The men move backward across the field, completing one entire section, tossing the unused plants into the unset field. Then reset the lines to cover another section. We were told that the usual day's work of transplanting, for a man under these conditions, after the field is fitted and the plants are brought to him, is two mow or one-third of an acre. The seven men in this group would thus set two and a third acres per day and, at the wage Mrs. Wu was paying, the cash outlay, if the help was hired, would be nearly 21 cents per acre. This is more cheaply than we are able to set cabbage and tobacco plants with our best machine methods. In Japan, as seen in Figs. 164 and 165, the women participate in the work of setting the plants more than in China.

After the rice has been transplanted its care, unlike that of our wheat crop, does not cease. It must be hoed, fertilized and watered. To facilitate the watering all fields have been leveled, canals, ditches and drains provided, and to aid in fertilizing and hoeing, the setting has been in rows and in hills in the row.

The first working of the rice fields after the transplanting, as we saw it in Japan, consisted in spading between the hills with a four-tined hoe, apparently more for loosening the soil and aeration than for killing weeds. After this treatment the field was gone over again in the manner seen in Fig. 166, where the man is using his bare hands to smooth and level the stirred soil, taking care to eradicate every weed, burying them beneath the mud, and to straighten each hill of rice as it is passed. Sometimes the fingers are armed with bamboo claws to facilitate the weeding. Machinery in the form of revolving hand cultivators is recently coming into use in Japan, and two men using these are seen in Fig. 14. In these

Fig. 164.—A group of Japanese women transplanting rice, in rainy weather costume, at Fukuoka Experiment Station, Japan.

cultivators the teeth are mounted on an axle so as to revolve as the cultivator is pushed along the row.

Fertilization for the rice crop receives the greatest attention everywhere by these three nations and in no direction more than in maintaining the store of organic matter in the soil. The pink clover, to which reference has been made, Figs. 99 and 100, is extensively sowed after a crop of rice is harvested in the fall and comes into full bloom, ready to cut for compost or to turn under directly when the rice fields are plowed. Eighteen to twenty tons of this green clover are produced per acre, and in Japan this is usually applied to about three acres, the stubble

Fig. 165.—Japanese young women transplanting rice, under broad sunshade hats.

and roots serving for the field producing the clover, thus giving a dressing of six to seven tons of green manure per acre, carrying not less than 37 pounds of potassium; 5 pounds of phosphorus, and 58 pounds of nitrogen.

Where the families are large and the holdings small, so they cannot spare room to grow the green manure crop, it is gathered on the mountain, weed and hill lands, or it may be cut in the canals. On our boat trip west from

Fig. 166.—Smoothing the soil and pulling weeds after the first working of a field of transplanted rice, Japan.

Soochow the last of May, many boats were passed carrying tons of the long green ribbon-like grass, cut and gathered from the bottom of the canal. To cut this grass men were working to their armpits in the water of the canal, using a crescent-shaped knife mounted like an anchor from the end of a 16-foot bamboo handle. This was shoved forward along the bottom of the canal and then drawn backward, cutting the grass, which rose to the surface where it was gathered upon the boats. Or material for green manure may be cut on grave, mountain or hill lands, as described under Fig. 115.

The straw of rice and other grain and the stems of any plant not usable as fuel may also be worked into the mud of rice fields, as may the chaff which is often scattered upon the water after the rice is transplanted, as in Fig. 168.

Reference has been made to the utilization of waste of various kinds in these countries to maintain the productive power of their soils, but it is worth while, in the interests of western nations, as helping them to realize the

Fig. 167.—Boat load of grass cut from bottom of canal, to be used as green manure or in preparing compost fertilizer, Kiangsu, China.

ultimate necessity of such economies, to state again, in more explicit terms, what Japan is doing. Dr. Kawaguchi, of the National Department of Agriculture and Commerce, taking his data from their records, informed me that Japan produced, in 1908, and applied to her fields, 23,850,295 tons of human manure; 22,812,787 tons of compost; and she imported 753,074 tons of commercial fertilizers, 7000 of which were phosphates in one form or another. In addition to these she must have applied not less than 1,404,000 tons of fuel ashes and 10,185,500 tons of green manure products grown on her hill and weed lands, and all of these applied to less than 14,000,000 acres of cultivated field, and it should be emphasized that this is done because as yet they have found no better way of permanently maintaining a fertility capable of feeding her millions.

Besides fertilizing, transplanting and weeding the rice crop there is the enormous task of irrigation to be maintained until the rice is nearly matured. Much of the water used is lifted by animal power and a large share of this is human. Fig. 169 shows two Chinese men in their

Fig. 168.—Applying chaff to a rice field as a fertilizer.

cool, capacious, nowhere-touching summer trousers flinging water with the swinging basket, and it is surprising the amount of water which may be raised three to four feet by this means. The portable spool windlass, in Figs. 27 and 123, has been described, and Fig. 170 shows the quadrangular, cone-shaped bucket and sweep extensively used in Chihli. This man was supplying water sufficient for the irrigation of half an acre, per day, lifting the water eight feet.

The form of pump most used in China and the foot-power for working it are seen in Fig. 171. Three men working a similar pump are seen in Fig. 150, a closer view of three men working the foot-power may be seen in Fig. 42 and still another stands adjacent to a series of flooded fields in Fig. 172. Where this view was taken the old farmer informed us that two men, with this pump, lifting water three feet, were able to cover two mow of land with three inches of water in two hours. This is at the

Fig. 169.—Irrigation by means of the swinging basket, Province of Chihli, China.

rate of 2.5 acre-inches of water per ten hours per man, and for 12 to 15 cents, our currency, thus making sixteen acre-inches, or the season's supply of water, cost 77 to 96 cents, where coolie labor is hired and fed. Such is the efficiency of human power applied to the Chinese pump, measured in American currency.

This pump is simply an open box trough in which travels a wooden chain carrying a series of loosely fitting boards which raise the water from the canal, discharging it into the field. The size of the trough and of the buckets are varied to suit the power applied and the amount of water to be lifted. Crude as it appears there is nothing in western manufacture that can compete with it in first cost, maintenance or efficiency for Chinese conditions and nothing is more characteristic of all these people than their efficient, simple appliances of all kinds, which they have reduced to the lowest terms in every feature of construction and cost. The greatest results are accomplished

Fig. 170.—Well sweep and quadrangular, conical water bucket used for irrigation in Chihli.

by the simplest means. If a canal must be bridged and it is too wide to be covered by a single span, the Chinese engineer may erect it at some convenient place and turn the canal under it when completed. This we saw in the case of a new railroad bridge near Sungkiang. The bridge was completed and the water had just been turned under it and was being compelled to make its own excavation. Great expense had been saved while traffic on the canal had not been obstructed.

In the foot-power wheel of Japan all gearing is eliminated and the man walks the paddles themselves, as seen in Fig. 173. Some of these wheels are ten feet in diameter, depending upon the hight the water must be lifted.

Irrigation by animal power is extensively practiced in each of the three countries, employing mostly the type of power wheel shown in Fig. 158. The next illustration, Fig. 174, shows the most common type of shelter seen in Chekiang and Kiangsu provinces, which are there very numer-

Fig. 171.—Three-man Chinese foot-power and wooden chain pump extensively used for irrigation in various parts of China.

ous. We counted as many as forty such shelters in a semicircle of half a mile radius. They provide comfort for the animals during both sunshine and rain, for under no conditions must the water be permitted to run low on the rice fields, and everywhere their domestic animals receive kind, thoughtful treatment.

In the less level sections, where streams have sufficient fall, current wheels are in common use, carrying buckets near their circumference arranged so as to fill when passing through the water, and to empty after reaching the highest level into a receptacle provided with a conduit which leads the water to the field. In Szechwan province some of these current wheels are so large and gracefully constructed as to strongly suggest Ferris wheels. A view of one of these we are permitted to present in Fig. 175, through the kindness of Rollin T. Chamberlin who took

Fig. 172.—Fields recently flooded with the Chinese foot-power chain pump preparatory to plowing for rice.

the photograph from which the engraving was prepared. This wheel which was some forty feet in diameter, was working when the snap shot was taken, raising the water and pouring it into the horizontal trough seen near the top of the wheel, carried at the summit of a pair of heavy poles standing on the far side of the wheel. From this trough, leading away to the left above the sky line, is the long pipe, consisting of bamboo stems joined together, for conveying the water to the fields.

When the harvest time has come, notwithstanding the large acreage of grain, yielding hundreds of millions of bushels, the small, widely scattered holdings and the surface of the fields render all of our machine methods quite impossible. Even our grain cradle, which preceded the reaper, would not do, and the great task is still met with the old time sickle, as seen in Fig. 176, cutting the rice hill by hill, as it was transplanted.

Fig. 173.—Japanese irrigation foot-wheel.

Previous to the time for cutting, after the seed is well matured, the water is drawn off and the land permitted to dry and harden. The rainy season is not yet over and much care must be exercised in curing the crop. The bundles may be shocked in rows along the margins of the paddies, as seen in Fig. 176, or they may be suspended, heads down, from bamboo poles as seen in Fig. 177.

The threshing is accomplished by drawing the heads of the rice through the teeth of a metal comb mounted as seen at the right in Fig. 178, near the lower corner, be-

Fig. 174.—Power-wheel shelter on bank of canal, in Kiangsu province, China.

Fig. 175.—Large current water-wheel in use in Szechwan province, China.
Photograph by Rollin T. Chamberlin.

hind the basket, where a man and woman are occupied in winnowing the dust and chaff from the grain by means of a large double fan. Fanning mills built on the principle of those used by our farmers and closely resembling them have long been used in both China and Japan. After the rice is threshed the grain must be hulled before it can serve as food, and the oldest and simplest method of polishing used by the Japanese is seen in Fig. 179, where the

Fig. 176.—Japanese farmers harvesting rice with the old-time sickle.

friction of the grain upon itself does the polishing. A quantity of rice is poured into the receptacle when, with heavy blows, the long-headed plunger is driven into the mass of rice, thus forcing the kernels to slide over one another until, by their abrasion, the desired result is secured. The same method of polishing, on a larger scale, is accomplished where the plungers are worked by the weight of the body, a series of men stepping upon lever handles of weighted plungers, raising them and allowing them to fall under the force of the weight attached. Recently, however, mills worked by gasoline engines are in operation for both hulling and polishing, in Japan.

The many uses to which rice straw is put in the economies of these people make it almost as important as the rice itself. As food and bedding for cattle and horses; as thatching material for dwellings and other shelters; as fuel; as a mulch; as a source of organic matter in the soil, and as a fertilizer, it represents a money value which

Fig. 177.—Suspending rice bundles from bamboo frames set up in the fields for curing the grain, preparatory to threshing, Japan.

is very large. Besides these ultimate uses the rice straw is extensively employed in the manufacture of articles used in enormous quantities. It is estimated that not less than 188,700,000 bags such as are seen in Figs. 180 and 181, worth $3,110,000 are made annually from the rice straw in Japan, for handling 346,150,000 bushels of cereals and 28,190,000 bushels of beans; and besides these, great numbers of bags are employed in transporting fish and other prepared manures.

In the prefecture of Hyogo, with 596 square miles of

farm land, as compared with Rhode Island's 712 square miles, Hyogo farmers produced in 1906, on 265,040 acres, 10,584,000 bushels of rice worth $16,191,400, securing an average yield of almost forty bushels per acre and a gross return of $61 for the grain alone. In addition to this, these farmers grew on the same land, the same season, at

Fig. 178.—Winnowing rice in Japan, using the large double fan worked by a pair of bamboo handles. A metal comb for removing the rice from the straw stands at the right.

least one other crop. Where this was barley the average yield exceeded twenty-six bushels per acre, worth $17.

In connection with their farm duties these Japanese families manufactured, from a portion of their rice straw, at night and during the leisure hours of winter, 8,980,000 pieces of matting and netting of different kinds having a market value of $262,000; 4,838,000 bags worth $185,000;

8,742,000 slippers worth \$34,000; 6,254,000 sandals worth \$30,000; and miscellaneous articles worth \$64,000. This is a gross earning of more than \$21,000,000 from eleven and a half townships of farm land and the labor of the

Fig. 179.—Large wooden mortar used for the polishing of rice in Japan.

farmers' families, an average earning of \$80 per acre on nearly three-fourths of the farm land of this prefecture. At this rate three of the four forties of our 160-acre farms should bring a gross annual income of \$9,600 and the fourth forty should pay the expenses.

At the Nara Experiment Station we were informed that

Fig. 180.—Sacking rice in bags made from the rice straw, Japan.

Fig. 181.—Loading, for shipment, rice put up in bags made from the rice straw, Japan.

the money value of a good crop of rice in that prefecture should be placed at ninety dollars per acre for the grain and eight dollars for the unmanufactured straw; thirty-six dollars per acre for the crop of naked barley and two dollars per acre for the straw. The farmers here practice a rotation of rice and barley covering four or five years, followed by a summer crop of melons, worth $320 per acre and some other vegetable instead of the rice on the

Fig. 182.—A Japanese family gathering and threshing barley, grown as a winter crop before rice.

fifth or sixth year, worth eighty yen per tan, or $160 per acre. To secure green manure for fertilizing, soy beans are planted each year in the space between the rows of barley, the barley being planted in November. One week after the barley is harvested the soy beans, which produce a yield of 160 kan per tan, or 5290 pounds per acre, are turned under and the ground fitted for rice, At these rates the Nara farmers are producing on four-fifths or five-sixths of their rice lands a gross earning of $136 per acre annually, and on the other fifth or sixth, an earning of $480 per acre, not counting the annual crop of soy beans used in maintaining the nitrogen and organic matter in their soils, and not counting their earnings from home manufactures. Can the farmers of our south Atlantic and Gulf Coast states, which are in the same latitude, sometime attain to this standard? We see no reason why they should not, but only with the best of irrigation, fertilization and proper rotation, with multiple cropping.

XIII.

SILK CULTURE.

Another of the great and in some ways one of the most remarkable industries of the Orient is that of silk production, and its manufacture into the most exquisite and beautiful fabrics in the world. Remarkable for its magnitude; for having had its birthplace apparently in oldest China, at least 2600 years B. C.; for having been founded on the domestication of a wild insect of the woods; and for having lived through more than four thousand years, expanding until a $1,000,000 cargo of the product has been laid down on our western coast at one time and rushed by special fast express to New York City for the Christmas trade.

Japan produced in 1907 26,072,000 pounds of raw silk from 17,154,000 bushels of cocoons, feeding the silkworms from mulberry leaves grown on 957,560 acres. At the export selling price of this silk in Japan the crop represents a money value of $124,000,000, or more than two dollars per capita for the entire population of the Empire; and engaged in the care of the silkworms, as seen in Figs. 184, 185, 186 and 187, there were, in 1906, 1,407,766 families or some 7,000,000 people.

Richard's geography of the Chinese Empire places the total export of raw silk to all countries, from China, in 1905, at 30,413,200 pounds, and this, at the Japanese export price, represents a value of $145,000,000. Richard also states that the value of the annual Chinese export of silk to France amounts to 10,000,000 pounds sterling and

that this is but twelve per cent of the total, from which it
appears that her total export alone reaches a value near
$400,000,000.

The use of silk in wearing apparel is more general among
the Chinese than among the Japanese, and with China's
eightfold greater population, the home consumption of silk
must be large indeed and her annual production must much

Fig. 184.—Removing silkworm eggs from sheets of paper where they were
laid, preparatory for hatching, Japan.

exceed that of Japan. Hosie places the output of raw
silk in Szechwan at 5,439,500 pounds, which is nearly a
quarter of the total output of Japan, and silk is extensively grown in eight other provinces, which together have
an area nearly fivefold that of Japan. It would appear,
therefore, that a low estimate of China's annual production of raw silk must be some 120,000,000 pounds, and
this, with the output of Japan and Korea, would make a
product for the three countries probably exceeding
150,000,000 pounds annually, representing a total value of
perhaps $700,000,000; quite equalling in value the wheat

crop of the United States, but produced on less than one-eighth of the area.

According to the observations of Count Dandola, the worms which contribute to this vast earning are so small that some 700,000 of them weigh at hatching only one pound, but they grow very rapidly, shed their skins four times, weighing 15 pounds at the time of the first moult,

Fig. 185.—Feeding silkworms. One of the 16 bamboo trays, on which the silk-worms are feeding, has been removed from the racks and Japanese girls are spreading over it a fresh supply of mulberry leaves.

94 pounds at the second, 400 pounds at the third, 1628 pounds at the fourth moulting and when mature have come to weigh nearly five tons—9500 pounds. But in making this growth during about thirty-six days, according to Paton, the 700,000 worms have eaten 105 pounds by the time of the first moult; 315 pounds by the second; 1050 pounds by the third; 3150 pounds by the fourth, and in the final period, before spinning, 19,215 pounds, thus consuming in all nearly twelve tons of mulberry leaves in producing nearly five tons of live weight, or at the rate

of two and a half pounds of green leaf to one pound of growth.

According to Paton, the cocoons from the 700,000 worms would weigh between 1400 and 2100 pounds and these, according to the observations of Hosie in the province of Szechwan, would yield about one-twelfth their weight of raw silk. On this basis the one pound of worms hatched from the eggs would yield between 116 and 175 pounds of

Fig. 186.—Providing places for silkworms to spin their cocoons.

raw silk, worth, at the Japanese export price for 1907, between $550 and $832, and 164 pounds of green mulberry leaves would be required to produce a pound of silk.

A Chinese banker in Chekiang province, with whom we talked, stated that the young worms which would hatch from the eggs spread on a sheet of paper twelve by eighteen inches would consume, in coming to maturity, 2660 pounds of mulberry leaves and would spin 21.6 pounds of silk. This is at the rate of 123 pounds of leaves to one pound of silk. The Japanese crop for 1907, 26,072,000 pounds, produced on 957,560 acres, is a mean yield of

27.23 pounds of raw silk per acre of mulberries, and this would require a mean yield of 4465 pounds of green mulberry leaves per acre, at the rate of 164 pounds per pound of silk.

Ordinary silk in these countries is produced largely from three varieties of mulberries, and from them there may be three pickings of leaves for the rearing of a spring, summer and autumn crop of silk. We learned at the Nagoya Experiment Station, Japan, that there good spring

Fig. 187.—Selecting the best cocoons, male and female determined by the shape and size, for purposes of breeding.

yields of mulberry leaves are at the rate of 400 kan, the second crop, 150 kan, and the third crop, 250 kan per tan, making a total yield of over thirteen tons of green leaves per acre. This, however, seems to be materially higher than the average for the Empire.

In Fig. 188 is a near view of a mulberry orchard in Chekiang province, which has been very heavily fertilized with canal mud, and which was at the stage for cutting the leaves to feed the first crop of silkworms. A bundle of cut limbs is in the crotch of the front tree in the view.

Those who raise mulberry leaves are not usually the feeders of the silkworms and the leaves from this orchard were being sold at one dollar, Mexican, per picul, or 32.25 cents per one hundred pounds. The same price was being paid a week later in the vicinity of Nanking, Kiangsu province.

The mulberry trees, as they appear before coming into leaf in the early spring, may be seen in Fig. 189. The long limbs are the shoots of the last year's growth, from which at least one crop of leaves had been picked, and in healthy orchards they may have a length of two to three

Fig. 188.—A near view of a mulberry orchard in Chekiang province, China.

feet. An orchard from a portion of which the limbs had just been cut, presented the appearance seen in Fig. 190. These trees were twelve to fifteen years old and the enlargements on the ends of the limbs resulted from the frequent pruning, year after year, at nearly the same place. The ground under these trees was thickly covered with a growth of pink clover just coming into bloom, which would be spaded into the soil, providing nitrogen and organic matter, whose decay would liberate potash, phosphorus and other mineral plant food elements for the crop.

In Fig. 191 three rows of mulberry trees, planted four

Fig. 189.—Near view of mulberry tree many years old, showing limbs of the last year's growth which will be cut close to the old wood when in full leaf.

feet apart, stand on a narrow embankment raised four feet, partly through adjusting the surrounding fields for rice, and partly by additions of canal mud used as a fertilizer. On either side of the mulberries is a crop of wind-

Fig. 190.—Mulberry orchard recently pruned for the first crop of leaves, with unpruned trees along the right.

sor beans, and on the left a crop of rape, both of which would be harvested in early June, the ground where they stand flooded, plowed and transplanted to rice. This and the other mulberry views were taken in the extensively canalized portion of China represented in Fig. 52. The farmer owning this orchard had just finished cutting two large bundles of limbs for the sale of the leaves in the village. He stated that his first crop ordinarily yields from three to as many as twenty piculs per mow, but that the second crop seldom exceeded two to three piculs. The first and second crop of leaves, if yielding together twenty-three piculs per mow, would amount to 9.2 tons per acre, worth, at the price named, $59.34. Mulberry leaves must be delivered fresh as soon as gathered and must be fed the same day, the limbs, when stripped of their leaves, at the place where these are sold, are tied into bundles and reserved for use as fuel.

In the south of China the mulberry is grown from low cuttings rooted by layering. We have before spoken of our five hours ride in the Canton delta region, on the steamer Nanning, through extensive fields of low mulberry then in full leaf, which were first mistaken for cotton nearing the blossom stage. This form of mulberry is seen in Fig. 43, and the same method of pruning is practiced in southern Japan. In middle Japan high pruning, as in Ckekiang and Kiangsu provinces, is followed, but in northern Japan the leaves are picked directly, as is the case with the last crop of leaves everywhere, pruning not being practiced in the more northern latitudes.

Not all silk produced in these northern countries is from the domesticated *Bombyx mori,* large amounts being obtained from the spinnings of wild silkworms feeding upon the leaves of species of oak growing on the mountain and hill lands in various parts of China, Korea and Japan. In China the collections in largest amount are reeled from the cocoons of the tussur worm (*Antheraea pernyi*) gathered in Shantung, Honan, Kweichow and Szechwan provinces. In the hilly parts of Manchuria also this industry

Fig. 191.—Three rows of mulberry trees occupying a long, narrow embankment which will be surrounded later by flooded rice fields.

is attaining large proportions, the cocoons being sent to Chefoo in the Shantung province, to be woven into pongee silk.

M. Randot has estimated the annual crop of wild silk cocoons in Szechwan at 10,180,000 pounds, although in the opinion of Alexander Hosie much of this may come from Kweichow. Richard places the export of raw wild silk from the whole of China proper, in 1904, at 4,400,000 pounds. This would mean not less than 75,300,000 pounds of wild cocoons and may be less than half the home consumption.

From data collected by Alexander Hosie it appears that in 1899 the export of raw tussur silk from Manchuria, through the port of Newchwang by steamer alone, was 1,862,448 pounds, valued at $1,721,200, and the production is increasing rapidly. The export from the same port the previous year, by steamer, was 1,046,704 pounds. This all comes from the hilly and mountain lands south of Mukden, lying between the Liao plain on the west and the Yalu river on the east, covering some five thousand square miles, which we crossed on the Antung-Mukden railway.

There are two broods of these wild silkworms each season, between early May and early October. Cocoons of the fall brood are kept through the winter and when the moths come forth they are caused to lay their eggs on pieces of cloth and when the worms are hatched they are fed until the first moult upon the succulent new oak leaves gathered from the hills, after which the worms are taken to the low oak growth on the hills where they feed themselves and spin their cocoons under the cover of leaves drawn about them.

The moths reserved from the first brood, after becoming fertile, are tied by means of threads to the oak bushes where they deposit the eggs which produce the second crop of tussur silk. To maintain an abundance of succulent leaves within reach the oaks are periodically cut back.

Thus these plain people, patient, frugal, unshrinking

from toil, the basic units of three of the oldest nations, go to the uncultivated hill lands and from the wild oak and the millions of insects which they help to feed upon it, not only create a valuable export trade but procure material for clothing, fuel, fertilizer and food, for the large chrysalides, cooked in the reeling of the silk, may be eaten at once or are seasoned with sauce to be used later. Besides this, the last unreelable portion of each cocoon is laid aside to be manufactured into silk wadding and into soft mattresses for caskets upon which the wealthy lay their dead.

XIV.

THE TEA INDUSTRY.

The cultivation of tea in China and Japan is another of the great industries of these nations, taking rank with that of sericulture, if not above it, in the important part it plays in the welfare of the people. There is little reason to doubt that the industry has its foundation in the need of something to render boiled water palatable for drinking purposes. The drinking of boiled water has been universally adopted in these countries as an individually available, thoroughly efficient and safe guard against that class of deadly disease germs which it has been almost impossible to exclude from the drinking water of any densely peopled country.

So far as may be judged from the success of the most thorough sanitary measures thus far instituted, and taking into consideration the inherent difficulties which must increase enormously with increasing populations, it appears inevitable that modern methods must ultimately fail in sanitary efficiency and that absolute safety must be secured in some manner having the equivalent effect of boiling water, long ago adopted by the Mongolian races, and which destroys active disease germs at the latest moment before using. And it must not be overlooked that the boiling of drinking water in China and Japan has been demanded quite as much because of congested rural populations as to guard against such dangers in large cities, while as yet our sanitary engineers have dealt only with the urban phases of this most vital problem and chiefly, too, thus far, only where it has been possible to procure the water supply in comparatively unpopulated hill lands. But such opportunities cannot remain available indefinitely, any more than they did in China and Japan, and already typhoid epidemics break out in our large cities and citizens are advised to boil their drinking water.

If tea drinking in the family is to remain general in most portions of the world, and especially if it shall increase in proportion to population, there is great industrial

Fig. 192.—Near view of tea garden with ground heavily mulched with straw, adjoining a Japanese farm village.

and commercial promise for China, Korea and Japan in their tea industry if they will develop tea culture still further over the extensive and still unused flanks of the hill lands; improve their cultural methods; their manufacture; and develop their export trade. They have the best of climatic and soil conditions and people sufficiently capable of enormously expanding the industry. Both improvement and expansion of methods along all essential lines, are needed, enabling them to put upon the market pure teas of thoroughly uniform grades of guaranteed quality, and with these the maintenance of an international code of rigid ethics which shall secure to all concerned a square deal and a fair division of the profits.

The production of rice, silk and tea are three industries which these nations are preeminently circumstanced and qualified to economically develop and maintain. Other nations may better specialize along other lines which fitness determines, and the time is coming when maximum production at minimum cost as the result of clean robust living that in every way is worth while, will determine lines of social progress and of international relations. With the vital awakening to the possibility of and necessity for world peace, it must be recognized that this can be nothing less than universal, industrial, commercial, intellectual and religious, in addition to making impossible forever the bloody carnage that has ravaged the world through all the centuries.

With the extension of rapid transportation and more rapid communication throughout the world, we are fast entering the state of social development which will treat the whole world as a mutually helpful, harmonious industrial unit. It must be recognized that in certain regions, because of peculiar fitness of soil, climate and people, needful products can be produced there better and enough more cheaply than elsewhere to pay the cost of transportation. If China, Korea and Japan, with parts of India, can and will produce the best and cheapest silks, teas or rice, it must be for the greatest good to seek a mutually helpful exchange, and the erection of impassable tariff barriers is a declaration of war and cannot make for world peace and world progress.

The date of the introduction of tea culture into China appears unknown. It was before the beginning of the Christian era and tradition would place it more than 2700 years earlier. The Japanese definitely date its introduction into their islands as in the year 805 A. D., and state its coming to them from China. However and whenever tea growing originated in these countries, it long ago attained and now maintains large proportions. In 1907

Japan had 124,482 acres of land occupied by tea gardens and tea plantations. These produced 60,877,975 pounds of cured tea, giving a mean yield of 489 pounds per acre. Of the more than sixty million pounds of tea produced annually on nearly two hundred square miles in Japan, less than twenty-two million pounds are consumed at home, the balance being exported at a cash value, in 1907, of $6,309,122, or a mean of sixteen cents per pound.

Fig. 193.—Looking across a tea plantation located on the flanks of wooded hill lands rising in the background, Japan.

In China the volume of tea produced annually is much larger than in Japan. Hosie places the annual export from Szechwan into Tibet alone at 40,000,000 pounds and this is produced largely in the mountainous portion of the province west of the Min river. Richard places her direct export to foreign countries, in 1905, at 176,027,255 pounds; and in 1906 at 180,271,000 pounds, so that the annual export must exceed 200,000,000 pounds, and her total product of cured tea must be more than 400,000,000.

The general appearance of tea bushes as they are grown in Japan is indicated in Fig. 192 The form of the bushes, the shape and size of the leaves and the dense green, shiny foliage quite suggests our box, so much used in borders and hedges. When the bushes are young, not covering the ground, other crops are grown between the rows, but as the

Fig. 194.—Group of Japanese women picking leaves of the tea plant.

bushes attain their full size, standing after trimming, waist to breast high, the ground between is usually thickly covered with straw, leaves or grass and weeds from the hill lands, which serve as a mulch, as a fertilizer, as a means of preventing washing on the hillsides, and to force the rain to enter the soil uniformly where it falls.

Quite a large per cent of the tea bushes are grown on small, scattering, irregular areas about dwellings, on land not readily tilled, but there are also many tea plantations of considerable size, presenting the appearance seen in Fig. 193. After each picking of the leaves the bushes are trimmed back with pruning shears, giving the rows the appearance of carefully trimmed hedges.

Fig. 195.—Weighing the freshly picked tea leaves in Japan.

The tea leaves are hand picked, generally by women and girls, after the manner seen in Fig. 194, where they are gathering the tender, newly-formed leaves into baskets to be weighed fresh, as seen in Fig. 195.

Three crops of leaves are usually gathered each season, the first yielding in Japan one hundred kan per tan, the second fifty kan and the third eighty kan per tan. This is at the rate of 3307 pounds, 1653 pounds, and 2645 pounds per acre, making a total of 7605 pounds for the season, from which the grower realizes from a little more than 2.2 to a little more than 3 cents per pound of the green leaves, or a gross earning of \$167 to \$209.50 per acre.

We were informed that the usual cost for fertilizers for the tea orchards was 15 to 20 yen per tan, or \$30 to \$40 per acre per annum, the fertilizer being applied in the fall, in the early spring and again after the first picking of the leaves. While the tea plants are yet small one winter crop and one summer crop of vegetables, beans or barley are grown between the rows, these giving a return of some forty dollars per acre. Where the plantations are given good care and ample fertilization the life of a plantation may be prolonged continuously, it is said, through one hundred or more years.

During our walk from Joji to Kowata, along a country road in one of the tea districts, we passed a tea-curing house. This was a long rectangular, one-story building with twenty furnaces arranged, each under an open window, around the sides. In front of each heated furnace with its tray of leaves, a Japanese man, wearing only a breech cloth, and in a state of profuse perspiration, was busy rolling the tea leaves between the palms of his hands.

At another place we witnessed the making of the low grade dust tea, which is prepared from the leaves of bushes which must be removed or from those of the prunings. In this case the dried bushes with their leaves were being beaten with flails on a threshing floor. The dust tea thus produced is consumed by the poorer people.

XV.

ABOUT TIENTSIN.

On the 6th of June we left central China for Tientsin and further north, sailing by coastwise steamer from Shanghai, again plowing through the turbid waters which give literal exactness to the name Yellow Sea. Our steamer touched at Tsingtao, taking on board a body of German troops, and again at Chefoo, and it was only between these two points that the sea was not strongly turbid. Nor was this all. From early morning of the 10th until we anchored at Tientsin, 2:30 P. M., our course up the winding Pei ho was against a strong dust-laden wind which left those who had kept to the deck as grey as though they had ridden by automobile through the Colorado desert; so the soils of high interior Asia are still spreading eastward by flood and by wind into the valleys and far over the coastal plains. Over large areas between Tientsin and Peking and at other points northward toward Mukden trees and shrubs have been systematically planted in rectangular hedgerow lines, to check the force of the winds and reduce the drifting of soils, planted fields occupying the spaces between.

It was on this trip that we met Dr. Evans of Shunking, Szechwan province. His wife is a physician practicing among the Chinese women, and in discussing the probable rate of increase of population among the Chinese, it was stated that she had learned through her practice that very many mothers had borne seven to eleven children and yet but one, two or at most three, were living. It was said there are many customs and practices which determine this high mortality among children, one of which is that of feeding them meat before they have teeth, the mother masticating for the children, with the result that often fatal convulsions follow. A Scotch physician of long experience in Shantung, who took the steamer at Tsingtao, replied to my question as to the usual size of families in his circuit, ''I do not know. It depends on the crops. In good years the number is large; in times of famine the girls especially are disposed of, often permitted to die when very young for lack of care. Many are sold at such times to go into other provinces.'' Such statements, however, should doubtless be taken with much allowance. If all the details were known regarding the cases which have served as foundations for such reports, the matter might appear in quite a different light from that suggested by such cold recitals.

Although land taxes are high in China Dr. Evans informed me that it is not infrequent for the same tax to be levied twice and even three times in one year. Inquiries regarding the land taxes among farmers in different parts of China showed rates running from three cents to a dollar and a half, Mexican, per mow; or from about eight cents to $3.87 gold, per acre. At these rates a forty acre farm would pay from $3.20 to $154.80, and a quarter section four times these amounts. Data collected by Consul-General E. T. Williams of Tientsin indicate that in Shantung the land tax is about one dollar per acre, and in Chihli, twenty cents. In Kiangsi province the rate is 200 to 300 cash per mow, and in Kiangsu, from 500 to 600 cash per mow, or, according to the rate of exchange given on page 76, from 60 to 80 cents, or 90 cents to $1.20 per acre in Kiangsi; and $1.50 to $2.00 or $1.80 to $2.40 in Kiangsu province. The lowest of these rates would make the land tax on 160 acres, $96, and the highest would place it at $384, gold.

In Japan the taxes are paid quarterly and the combined amount of the national, prefectural and village assessments usually aggregates about ten per cent of the government valuation placed on the land. The mean valuation placed on the irrigated fields, excluding Formosa and Karafuto, was in 1907, 35.35 yen per tan; that of the upland fields, 9.40 yen, and the *genya* and pasture lands were given a valuation of .22 yen per tan. These are valuations of $70.70, $18.80 and $.44, gold, per acre, respectively, and the taxes on forty acres of paddy field would be $282.80; $75.20 on forty acres of upland field, and $1.76, gold, on the same area of the *genya* and weed lands.

In the villages, where work of one or another kind is done for pay, Dr. Evans stated that a woman's wage might not exceed $8, Mexican, or $3.44, gold, per year, and when we asked how it could be worth a woman's while to work a whole year for so small a sum, his reply was, "If she did not do this she would earn nothing, and this would keep her in clothes and a little more." A cotton spinner in his church would procure a pound of cotton and on returning the yarn would receive one and a quarter pounds of cotton in exchange, the quarter pound being her compensation.

Dr. Evans also described a method of rooting slips from trees, practiced in various parts of China. The under side of a branch is cut, bent upward and split for a short distance; about this is packed a ball of moistened earth wrapped in straw to retain the soil and to provide for future watering; the whole may then be bound with strips of bamboo for greater stability. In this way slips for new mulberry orchards are procured.

At eight o'clock in the morning we entered the mouth of the Pei ho and wound westward through a vast, nearly sea-level, desert plain and in both directions, far toward the horizon, huge white stacks of salt dotted the surface of the Taku Government salt fields, and revolving in the wind were great numbers of horizontal sail windmills, pumping sea water into an enormous acreage of evaporation basins. In Fig. 196 may be seen five of the large salt stacks and six of the wind-mills, together with many smaller piles of salt. Fig. 197 is a closer view of the evaporation basins with piles of salt scraped from the surface after the mother liquor had been drained away. The windmills, which were working one, sometimes two, of the large wooden chain pumps, were some thirty feet in diameter and lifted the brine from tide-water basins into those of a second and third higher level where the second and final concentration occurred. These windmills, crude as they appear in Fig. 198, are nevertheless efficient, cheaply constructed and easily controlled. The eight sails, each six by ten feet, were so hung as to take the wind through the entire revolution, tilting automatically to receive the wind on the opposite face the moment the edge passed the critical point. Some 480 feet of sail surface were thus spread to the wind, working on a radius of fifteen feet. The horizontal drive wheel had a diameter of ten feet, carried eighty-eight wooden cogs which engaged a pinion with fifteen leaves, and there were nine arms on the reel at the other end of the shaft which drove the chain. The boards or buckets of the chain pump were six by twelve inches, placed nine inches apart, and with a fair breeze the pump ran full.

Fig. 196.—Salt stacks and sail windmills on the Taku evaporation fields at the mouth of the Pei ho, Chihli.

Enormous quantities of salt are thus cheaply manufactured through wind, tide and sun power directed by the cheapest human labor. Before reaching Tientsin we passed the Government storage yards and counted two hundred stacks of salt piled in the open, and more than a third of the yard had been passed before beginning the count. The average content of each stack must have exceeded 3000

Fig. 197.—Near view of evaporating basins with piles of salt ready to be removed from the fields.

cubic feet of salt, and more than 40,000,000 pounds must have been stored in the yards. Armed guards in military uniform patrolled the alleyways day and night. Long strips of matting laid over the stacks were the only shelter against rain.

Throughout the length of China's seacoast, from as far north as beyond Shanhaikwan, south to Canton, salt is manufactured from sea water in suitable places. In Szechwan province, we learn from the report of Consul-General Hosie, that not less than 300,000 tons of salt are annually manufactured there, largely from brine raised by animal power from wells seven hundred to more than two thousand feet deep.

Hosie describes the operations at a well more than two thousand feet deep, at Tzeliutsing. In the basement of a power-house which sheltered forty water buffaloes, a huge

Fig. 198.—Sail windmill used in pumping brine at the Taku Government salt works, Chihli, China.

bamboo drum twelve feet high, sixty feet in circumference, was so set as to revolve on a vertical axis propelled by four cattle drawing from its circumference. A hemp rope was wound about this drum, six feet from the ground, passing out and under a pulley at the well, then up and around a wheel mounted sixty feet above and descended to the bucket made from bamboo stems four inches in diameter and nearly sixty feet long, which dropped with great speed to the bottom of the well as the rope unwound. When the bucket reached the bottom four attendants, each with a buffalo in readiness, hitched to the drum and drove at a running pace, during fifteen minutes, or until the bucket was raised from the well. The buffalo were then unhitched and, while the bucket was being emptied and again dropped to the bottom of the well, a fresh relay were brought to the drum. In this way the work continued night and day.

The brine, after being raised from the well, was emptied into distributing reservoirs, flowing thence through bamboo pipes to the evaporating sheds where round bottomed, shallow iron kettles four feet across were set in brick arches in which jets of natural gas were burning.

Within an area some sixty miles square there are more than a thousand brine and twenty fire wells from which fuel gas is taken. The mouths of the fire wells are closed with masonry, out from which bamboo conduits coated with lime lead to the various furnaces, terminating with iron burners beneath the kettles. Remarkable is the fact that in the city of Tzeliutsing, both these brine and the fire wells have been operated in the manufacture of salt since before Christ was born.

The forty water buffalo are worth $30 to $40 per head and their food fifteen to twenty cents per day. The cost of manufacturing this salt is placed at thirteen to fourteen cash per catty, to which the Government adds a tax of nine cash more, making the cost at the factory from 82 cents to $1.15, gold, per hundred pounds. Salt manufacture is a Government monopoly and the product must be sold either to Government officials or to merchants who have bought the exclusive right to supply certain districts. The importation of salt is prohibited by treaties. For the salt tax collection China is divided into eleven circuits each having its own source of supply and transfer of salt from one circuit to another is forbidden.

The usual cost of salt is said to vary between one and a half and four cash per catty. The retail price of salt

ranges from three-fourths to three cents per pound, fully
twelve to fifteen times the cost of manufacture. The
annual production of salt in the Empire is some 1,860,000
tons, and in 1901 salt paid a tax close to ten million
dollars.

Beyond the salt fields, toward Tientsin, the banks of
the river were dotted at short intervals with groups of
low, almost windowless houses, Fig. 199, built of earth
brick plastered with clay on sides and roof, made more

Fig. 199.—Chinese village on the bank of the Pei ho, Province of Chihli.

resistant to rain by an admixture of chaff and cut straw,
and there was a remarkable freshness of look about them
which we learned was the result of recent preparations
made for the rainy season about to open. Beyond the
first of these villages came a stretch of plain dotted thickly
and far with innumerable grave mounds, to which reference has been made. For nearly an hour we had traveled
up the river before there was any material vegetation, the
soil being too saline apparently to permit growth, but
beyond this, crops in the fields and gardens, with some
fruit and other trees, formed a fringe of varying width

along the banks. Small fields of transplanted rice on both banks were frequent and often the land was laid out in beds of two levels, carefully graded, the rice occupying the lower areas, and wooden chain pumps were being worked by hand, foot and animal power, irrigating both rice and garden crops.

In the villages were many stacks of earth compost, of the Shantung type; manure middens were common and donkeys drawing heavy stone rollers followed by men with large wooden mallets, were going round and round, pulverizing and mixing the dry earth compost and the large earthen brick from dismantled kangs, preparing fertilizer for the new series of crops about to be planted, following the harvest of wheat and barley. Large boatloads of these prepared fertilizers were moving on the river and up the canals to the fields.

Toward the coast from Tientsin, especially in the country traversed by the railroad, there was little produced except a short grass, this being grazed at the time of our visit and, in places, cut for a very meagre crop of hay. The productive cultivated lands lie chiefly along the rivers and canals or other water courses, where there is better drainage as well as water for irrigation. The extensive, close canalization that characterizes parts of Kiangsu and Chekiang provinces is lacking here and for this reason, in part, the soil is not so productive. The fuller canalization, the securing of adequate drainage and the gaining of complete control of the flood waters which flow through this vast plain during the rainy season constitute one of China's most important industrial problems which, when properly solved, must vastly increase her resources. During our drive over the old Peking-Taku road saline deposits were frequently observed which had been brought to the surface during the dry season, and the city engineer of Tientsin stated that in their efforts at parking portions of the foreign concessions they had found the trees dying after a few years when their roots began to penetrate the more saline subsoil, but that since they had opened canals, improving the drainage, trees were no longer dying. There is little doubt that proper drainage by means of canals, and the irrigation which would go with it, would make all of these lands, now more or less saline, highly productive, as are now those contiguous to the existing water courses.

Fig. 200.—China's method of shallow cultivation, producing an earth mulch to conserve soil moisture.

It had rained two days before our drive over the Taku road and when we applied for a conveyance the proprietor doubted whether the roads were passible, as he had been compelled to send out an extra team to assist in the return of one which had been stalled during the previous night. It was finally arranged to send an extra horse with us. The rainy season had just begun but the deep trenching of the roads concentrates the water in them and greatly intensifies the trouble. In one of the little hamlets through which we passed the roadway was trenched to a depth of three to four feet in the middle of the narrow street, leaving only five feet for passing in front of the dwellings on either side, and in this trench our carriage moved through mud and water nearly to the hubs.

Between Tientsin and Peking, in the early morning after a rain of the night before, we saw many farmers working their fields with the broad hoes, developing an earth mulch at the first possible moment to conserve their much needed moisture. Men were at work, as seen in Figs. 200 and 201, using long handled hoes, with blades

Fig. 201.—Hoe used for shallow cultivation in developing an earth mulch. The blade is 13 inches long and 9 inches wide.

nine by thirteen inches, hung so as to draw just under the surface, doing very effective work, permitting them to cover the ground rapidly.

Walking further, we came upon six women in a field of wheat, gleaning the single heads which had prematurely ripened and broken over upon the ground between the rows soon to be harvested. Whether they were doing this as a privilege or as a task we do not know; they were strong, cheerful, reasonably dressed, hardly past middle life and it was nearly noon, yet not one of them had collected more straws than she could readily grasp in one hand. The season in Chihli as in Shantung, had been one of unusual drought, making the crop short and perhaps unusual frugality was being practiced; but it is in saving that these people excel perhaps more than in producing. These heads of wheat, if left upon the ground, would be wasted and if the women were privileged gleaners in the fields their returns were certainly much greater than were

Fig. 202.—Gathering wheat, harvested by being pulled and tied in bundles. Team consists of a small donkey and a medium sized cow, which constitute the most common farm team. Tientsin, China.

those of the very old women we have seen in France gathering heads of wheat from the already harvested fields.

In the fields between Tientsin and Peking all wheat was being pulled, the earth shaken from the roots, tied in small bundles and taken to the dwellings, sometimes on the heavy cart drawn by a team consisting of a small donkey and cow hitched tandem, as seen in Fig. 202. Millet had been planted between the rows of wheat in this field and was already up. When the wheat was removed the ground would be fertilized and planted to soy beans. Because of the dry season this farmer estimated his yield would be but eight to nine bushels per acre. He was expecting to harvest thirteen to fourteen bushels of millet and from ten to twelve bushels of soy beans per acre from the same field. This would give him an earning, based on the local prices, of $10.36, gold, for the wheat; $6.00 for the beans, and $5.48 per acre for the millet. This land was owned by the family of the Emperor and was rented at $1.55, gold, per acre. The soil was a rather light sandy loam, not inherently fertile, and fertilizers to the value of $3.61 gold, per acre, had been applied, leaving the earning $16.71 per acre.

Another farmer with whom we talked, pulling his crop of wheat, would follow this with millet and soy beans in alternate rows. His yield of wheat was expected to be eleven to twelve bushels per acre, his beans twenty-one bushels and his millet twenty-five bushels which, at the local prices for grain and straw, would bring a gross earning of $35, gold, per acre.

Before reaching the end of our walk through the fields toward the next station we came across another of the many instances of the labor these people are willing to perform for only a small possible increase in crop. The field was adjacent to one of the windbreak hedges and the trees had spread their roots far afield and were threatening his crop through the consumption of moisture and plant food. To check this depletion the farmer had dug a trench twenty inches deep the length of his field, and some twenty feet from the line of trees, thereby cutting all of the surface roots to stop their draft on the soil. The trench was left open and an interesting feature observed was that nearly every cut root on the field side of the trench had thrown up one or more shoots bearing leaves, while the ends still connected with the trees showed no signs of leaf growth.

In Chihli as elsewhere the Chinese are skilled gardeners, using water for irrigation whenever it is advantageous. One gardener was growing a crop of early cabbage, followed by one of melons, and these with radish the same season. He was paying a rent of $6.45, gold, per acre; was applying fertilizer at a cost of nearly $8 per acre for each of the three crops, making his cash outlay $29.67 per acre. His crop of cabbage sold for $103, gold; his melons for $77, and his radish for something more than $51, making a total of $232.20 per acre, leaving him a net value of $202.53.

A second gardener, growing potatoes, obtained a yield, when sold new, of 8,000 pounds per acre; and of 16,000 pounds when the crop was permitted to mature. The new potatoes were sold so as to bring $51.60 and the mature potatoes $185.76 per acre, making the earning for the two crops the same season a total of $237.36, gold. By planting the first crop very early these gardeners secure two crops the same season, as far north as Columbus, Ohio, and Springfield, Illinois, the first crop being harvested when the tubers are about the size of walnuts. The rental and fertilizers in this case amounted to $30.96 per acre.

Still another gardener growing winter wheat followed by onions, and these by cabbage, both transplanted, realized from the three crops a gross earning of $176.73, gold, per acre, and incurred an expense of $31.73 per acre for fertilizer and rent, leaving him a net earning of $145 per acre.

These old people have acquired the skill and practice of storing and preserving such perishable fruits as pears and grapes so as to enable them to keep them on the markets almost continuously. Pears were very common in the latter part of June, and Consul-General Williams informed me that grapes are regularly carried into July. In talking with my interpreter as to the methods employed I could only learn that the growers depend simply upon dry earth cellars which can be maintained at a very uniform temperature, the separate fruits being wrapped in paper. No foreigner with whom we talked knew their methods.

Vegetables are carried through the winter in such earth cellars as are seen in Fig. 88, page 161, these being covered after they are filled.

As to the price of labor in this part of China, we learned through Consul-General Williams that a master mechanic may receive 50 cents, Mexican, per day, and a journeyman 18 cents, or at a rate of 21.5 cents and 7.75 cents, gold. Farm laborers receive from $20 to $30, Mexican, or $8.60 to $12.90, gold, per year, with food, fuel and presents which make a total of $17.20 to $21.50. This is less for the year than we pay for a month of probably less efficient labor. There is relatively little child labor in China and this perhaps should be expected when adult labor is so abundant and so cheap.

XVI.

MANCHURIA AND KOREA.

The 39th parallel of latitude lies just south of Tientsin; followed westward, it crosses the toe of Italy's boot, leads past Lisbon in Portugal, near Washington and St. Louis and to the north of Sacramento on the Pacific. We were leaving a country with a mean July temperature of 80° F., and of 21° in January, but where two feet of ice may form; a country where the eighteen year mean maximum temperature is 103.5° and the mean minimum 4.5°; where twice in this period the thermometer recorded 113° above zero, and twice 7° below, and yet near the coast and in the latitude of Washington; a country where the mean annual rainfall is 19.72 inches and all but 3.37 inches falls in June, July, August and September. We had taken the 5:40 A. M. Imperial North-China train, June 17th, to go as far northward as Chicago,—to Mukden in Manchuria, a distance by rail of some four hundred miles, but all of the way still across the northward extension of the great Chinese coastal plain. Southward, out from the coldest quarter of the globe, where the mean January temperature is more than 40° below zero, sweep northerly winds which bring to Mukden a mean January temperature only 3° above zero, and yet there the July temperature averages as high as 77° and there is a mean annual rainfall of but 18.5 inches, coming mostly in the summer, as at Tientsin.

Although the rainfall of the northern extension of China's coastal plain is small, its efficiency is relatively high because of its most favorable distribution and the high summer temperatures. In the period of early growth, April, May and June, there are 4.18 inches; but in the period of maximum growth, July and August, the rainfall is 11.4 inches; and in the ripening period, September and October, it is 3.08 inches, while during the rest of the year but 1.06 inch falls. Thus most of the rain comes at the time when the crops require the greatest daily consumption and it is least in mid-winter, during the period of little growth.

As our train left Tientsin we traveled for a long distance through a country agriculturally poor and little tilled, with surface flat, the soil apparently saline, and the land greatly in need of drainage. Wherever there were canals the crops were best, apparently occupying more or less continuous areas along either bank. The day was hot and sultry but laborers were busy with their large hoes, often with all garments laid aside except a short shirt or a pair of roomy trousers.

In the salt district about the village of Tangku there were huge stacks of salt and smaller piles not yet brought together, with numerous windmills, constituting most striking features in the landscape, but there was almost no agricultural or other vegetation. Beyond Pehtang there are other salt works and a canal leads westward to Tientsin, on which the salt is probably taken thither, and still other salt stacks and windmills continued visible until near Hanku, where another canal leads toward Peking. Here the coast recedes eastward from the railway and beyond the city limits many grave mounds dot the surrounding plains where herds of sheep were grazing.

As we hurried toward the delta region of the Lwan ho, and before reaching Tangshan, a more productive country was traversed. Thrifty trees made the landscape green, and fields of millet, kaoliang and wheat stretched for miles together along the track and back over the flat plain beyond the limit of vision. Then came fields planted with two rows of maize alternating with one row of soy beans, but not over twenty-eight inches apart, one stalk of corn in a place every sixteen to eighteen inches, all carefully hoed, weedless and blanketed with an excellent earth mulch; but still the leaves were curling in the intense heat of the sun. Tangshan is a large city, apparently of recent growth on the railroad in a country where isolated conical hills rise one hundred or two hundred feet out of the flat plains. Cart loads of finely pulverized earth compost were here moving to the fields in large numbers, being laid in single piles of five hundred to eight hundred pounds, forty to sixty feet apart. At Kaiping the country grows a little rolling and we passed through the first railway cuts, six to eight feet deep, and the water in the streams is running ten to twelve feet below the surface of the fields. On the right and beyond Kuyeh there are low hills, and here we passed enormous quantities of dry, finely powdered earth compost, distributed on narrow unplanted area over the fields. What crop, if indeed any, had occupied these areas this season, we could not judge. The fertilization here is even more extensive and more general than we found it in the Shantung province, and in places water was being carried in pails to the fields for use either in planting or in transplanting, to ensure the readiness of the new crops to utilize the first rainfall when it comes.

Then the bed of a nearly dry stream some three hundred feet wide was crossed and beyond it a sandy plain was planted in long narrow fields between windbreak hedges. The crops were small but evidently improved by the influence of the shelter. The sand in places had drifted into the hedges to a hight of three feet. At a number of other places along the way before Mukden was reached such protected areas were passed and oftenest on the north side of wide, now nearly dry, stream channels.

As we passed on toward Shanhaikwan we were carried over broad plains even more nearly level and unobstructed than any to be found in the corn belt of the middle west, and these too planted with corn, kaoliang, wheat and beans, and with the low houses hidden in distant scattered clusters of trees dotting the wide plain on either side, with not a fence, and nothing to suggest a road anywhere in sight. We seemed to be moving through one vast field dotted with hundreds of busy men, a plowman here, and there a great cart hopelessly lost in the field so far as one could see any sign of road to guide their course.

Fig. 203.—Exportation of soy beans from Manchuria. Lwanchow, Chihli.

Some early crop appeared to have been harvested from areas alternating with those on the ground, and these were dotted with piles of the soil and manure compost, aggregating hundreds of tons, distributed over the fields but no doubt during the next three or four days these thousands of piles would have been worked into the soil and vanished from sight, to reappear after another crop and another year.

It was at Lwanchow that we met the out-going tide of soy beans destined for Japan and Europe, pouring in from the surrounding country in gunny sacks brought on heavy carts drawn by large mules, as seen in Fig. 203, and enormous quantities had been stacked in the open along the tracks, with no shelter whatever, awaiting the arrival of trains to move them to export harbors.

The planting here, as elsewhere, is in rows, but not of one kind of grain. Most frequently two rows of maize, kaoliang or millet alternated with the soy beans and usually not more than twenty-eight inches apart, sharp high ridge cultivation being the general practice. Such planting secures the requisite sunshine with a larger number of plants on the field; it secures a continuous general distribution of the roots of the nitrogen-fixing soy beans in the soil of all the field every season, and permits the soil to be more continuously and more completely laid under tribute by the root systems. In places where the stand of corn or millet was too open the gaps were filled with the soy beans. Such a system of planting possibly permits a more immediate utilization of the nitrogen gathered from the soil air in the root nodules, as these die and undergo nitrification during the same season, while the crops are yet on the ground, and so far as phosphorus and potassium compounds are liberated by this decay, they too would become available to the crops.

The end of the day's journey was at Shanhaikwan on the boundary between Chihli and Manchuria, the train stopping at 6:20 P. M. for the night. Stepping upon the veranda from our room on the second floor of a Japanese inn in the early morning, there stood before us, sullen and grey, the eastern terminus of the Great Wall, winding fifteen hundred miles westward across twenty degrees of longitude, having endured through twenty-one centuries, the most stupendous piece of construction ever conceived by man and executed by a nation. More than twenty feet thick at the base and than twelve feet on the top; rising fifteen to thirty feet above the ground with parapets along both faces and towers every two hundred yards rising twenty feet higher, it must have been, for its time and the methods of warfare then practiced, when defended by their thousands, the boldest and most efficient national defense ever constructed. Nor in the economy of construction and maintenance has it ever been equalled.

Even if it be true that 20,000 masons toiled through ten years in its building, defended by 400,000 soldiers, fed by a commissariat of 20,000 more and supported by 30,000 others in the transport, quarry and potters' service, she would then have been using less than eight tenths per cent of her population, on a basis of 60,000,000 at the time; while according to Edmond Théry's estimate, the officers and soldiers of Europe today, in time of peace, constitute one per cent of a population of 400,000,000 of people, and these, at only one dollar each per day for food, clothing and loss of producing power would cost her nations, in ten years, more than $14,000 million. China, with her present habits and customs, would more easily have maintained her army of 470,000 men on thirty cents each per day, or for a total ten-year cost of but $520,000,000. The French cabinet in 1900 approved a naval program involving an expenditure of $600,000,000 during the next ten years, a tax of more than $15 for every man, woman and child in the Republic.

Leaving Shanhaikwan at 5:20 in the morning and reaching Mukden at 6-30 in the evening, we rode the entire day through Manchurian fields. Manchuria has an area of 363,700 square miles, equal to that of both Dakotas, Minnesota, Nebraska and Iowa combined. It has roughly the outline of a huge boot and could one slide it eastward until Port Arthur was at Washington, Shanhaikwan would fall well toward Pittsburg, both at the tip of the broad toe to the boot. The foot would lie across Pennsylvania, New York, New Jersey and all of New England, extending beyond New Brunswick with the heel in the Gulf of St. Lawrence. Harbin, at the instep of the boot, would lie fifty miles east of Montreal and the expanding leg would reach northwestward nearly to James Bay, entirely to the north of the Ottawa river and the Canadian Pacific, spanning a thousand miles of latitude and nine hundred miles of longitude.

The Liao plain, thirty miles wide, and the central Sungari plain, are the largest in Manchuria, forming together a long narrow valley floor between two parallel mountain systems and extending northeasterly from the Liao gulf, between Port Arthur and Shanhaikwan, up the Liao river and down the Sungari to the Amur, a distance of eight hundred or more miles. These plains have a fertile, deep soil and it is on them and other lesser river bottoms that Manchurian agriculture is developed, supporting eight or nine million people on a cultivated acreage possibly not greater than 25,000 square miles.

Manchuria has great forest and grazing possibilities awaiting future development, as well as much mineral wealth. The population of Tsitsihar, in the latitude of middle North Dakota, swells from thirty thousand to seventy thousand during September and October, when the Mongols bring in their cattle to market. In the middle province, at the head of steam navigation on the Sungari, because of the abundance and cheapness of lumber, Kirin has become a ship-building center for Chinese junks. The Sungari—Milky—river, is a large stream carrying more water at flood season than the Amur above its mouth, the latter being navigable 450 miles for steamers drawing twelve feet of water, and 1500 miles for those drawing four feet, so that during the summer season the middle and northern provinces have natural inland waterways, but the outlet to the sea is far to the north and closed by ice six months of the year.

Not far beyond the Great Wall of China, fast falling into ruin, partly through the appropriation of its material for building purposes now that it has outlived its usefulness, another broad, nearly dry stream bed was crossed. There, in full bloom, was what appeared to be the wild white rose seen earlier, further south, west of Suchow, having a remarkable profusion of small white bloom in clusters resembling the Rambler rose. One of these bushes growing wild there on the bank of the canal had overspread a clump of trees one of which was thirty feet in hight, enveloping it in a mantle of bloom, as seen in the upper section of Fig. 204. The lower section of the illus-

Fig. 204.—Wild white rose in bloom west of Suchow, June 2d, and in southern Manchuria, June 18th. Lower section, close view of same, showing clusters.

tration is a closer view showing the clusters. The stem of this rose, three feet above the ground, measured 14.5 inches in circumference. If it would thrive in this country nothing could be better for parks and pleasure drives.

Later on our journey we saw it many times in bloom along the railway between Mukden and Antung, but nowhere attaining so large growth. The blossoms are scant three-fourths inch in diameter, usually in compact clusters of three to eleven, sometimes in twos and occasionally standing singly. The leaves are five-foliate, sometimes trifoliate; leaflets broadly lanceolate, accuminate and finely serrate; thorns minute, recurrent and few, only on the smaller branches.

In a field beyond, a small donkey was drawing a stone roller three feet long and one foot in diameter, firming the crests of narrow, sharp, recently formed ridges, two at a time. Millet, maize and kaoliang were here the chief crops. Another nearly dry stream was crossed, where the fields became more rolling and much cut by deep gulleys, the first instances we had seen in China except on the steep hillsides about Tsingtao. Not all of the lands here were cultivated, and on the untilled areas herds of fifty to a hundred goats, pigs, cattle, horses and donkeys were grazing.

Fields in Manchuria are larger than in China and some rows were a full quarter of a mile long, so that cultivation was being done with donkeys and cattle, and large numbers of men were working in gangs of four, seven, ten, twenty, and in one field as high as fifty, hoeing millet. Such a crew as the largest mentioned could probably be hired at ten cents each, gold, per day, and were probably men from the thickly settled portions of Shantung who had left in the spring, expecting to return in September or October. Both laborers and working animals were taking dinner in the fields, and earlier in the day we had seen several instances where hay and feed were being taken to the field on a wooden sled, with the plow and other tools. At noon this was serving as manger for the cattle, mules or donkeys.

In fields where the close, deep furrowing and ridging was being done the team often consisted of a heavy ox

and two small donkeys driven abreast, the three walking
in adjacent rows, the plow following the ox, or a heavy
mule instead.

The rainy season had not begun and in many fields there
was planting and transplanting where water was used
in separate hills, sometimes brought in pails from a near-
by stream, and in other cases on carts provided with tanks.
Holes were made along the crests of the ridges with the
blade of a narrow hoe and a little water poured in each
hill, from a dipper, before planting or setting.   These must
have been other instances where the farmers were willing
to incur additional labor to save time for the maturing of
the crop by assisting germination in a soil too dry to make
it certain until the rains came.

It appears probable that the strong ridging and the
close level rows so largely adopted here must have marked
advantages in utilizing the rainfall, especially the portions
coming early, and that later also if it should come in heavy
showers.   With steep narrow ridging, heavy rains would
be shed at once to the bottom of the deep furrows without
over-saturating the ridges, while the wet soil in the bottom of the furrows would favor deep percolation with
lateral capillary flow taking place strongly under the
ridges from the furrows, carrying both moisture and soluble plant food where they will be most completely and
quickly available.   When the rain comes in heavy showers
each furrow may serve as a long reservoir which will
prevent washing and at the same time permit quick penetration; the ridges never becoming flooded or puddled,
permit the soil air to escape readily as the water from the
furrows sinks, as it cannot easily do in flat fields when
the rains fall rapidly and fill all of the soil pores, thus
closing them to the escape of air from below, which must
take place before the water can enter.

When rows are only twenty-four to twenty-eight inches
apart, ridging is not sufficiently more wasteful of soil
moisture, through greater evaporation because of increased
surface, to compensate for the other advantages gained,

and hence their practice, for their conditions, appears sound.

The application of finely pulverized earth compost to fields to be planted, and in some cases where the fields were already planted, continued general after leaving Shanhailkwan as it had been before. Compost stacks were common in yards wherever buildings were close enough to the track to be seen. Much of the way about one-third of the fields were yet to be, or had just been, planted and in a great majority of these compost fertilizer had been laid down for use on them, or was being taken to them in large heavy carts drawn sometimes by three mules. Between Sarhougon and Ningyuenchow fourteen fields thus fertilized were counted in less than half a mile; ten others in the next mile; eleven in the mile and a quarter following. In the next two miles one hundred fields were counted and just before reaching the station we counted during five minutes, with watch in hand, ninety-five fields to be planted, upon which this fertilizer had been brought. In some cases the compost was being spread in furrows between the rows of a last year's crop, evidently to be turned under, thus reversing the position of the ridges.

After passing Lienshan, where the railway runs near the sea, a sail was visible on the bay and many stacks of salt piled about the evaporation fields were associated with the revolving sail windmills already described. Here, too, large numbers of cattle, horses, mules and donkeys were grazing on the untilled low lands, beyond which we traversed a section where all fields were planted, where no fertilizer was piled in the field but where many groups of men were busy hoeing, sometimes twenty in a gang.

Chinese soldiers with bayonetted guns stood guard at every railway station between Shanhaikwan and Mukden, and from Chinchowfu our coach was occupied by some Chinese official with guests and military attendants, including armed soldiers. The official and his guests were an attractive group of men with pleasant faces and winning manners, clad in many garments of richly figured silk of bright, attractive, but unobtrusive, colors, who talked, seriously or in mirth, almost incessantly. They took the train about one o'clock and lunch was immediately served in Chinese style, but the last course was not brought until nearly four o'clock. At every station soldiers stood in line in the attitude of salute until the official car had passed.

Just before reaching Chinchowfu we saw the first planted fields littered with stubble of the previous crop, and in many instances such stubble was being gathered and removed to the villages, large stacks having been piled in the yards to be used either as fuel or in the production of compost. As the train approached Taling ho groups of men were hoeing in millet fields, thirty in one group on one side and fifty in another body on the other. Many small herds of cattle, horses, donkeys and flocks of goats and sheep were feeding along stream courses and on the unplanted fields. Beyond the station, after crossing the river, still another sand dune tract was passed, planted with willows, millet occupying the level areas between the dunes, and not far beyond, wide untilled flats were crossed, on which many herds were grazing and dotted with grave mounds as we neared Koupantze, where a branch of the railway traverses the Liao plain to the port of Newchwang. It was in this region that there came the first suggestion of resemblance to our marshland meadows; and very soon there were seen approaching from the distance loads so green that except for the large size one would have judged them to be fresh grass. They were loads of cured hay in the brightest green, the result, no doubt, of curing under their dry weather conditions.

At Ta Hu Shan large quantities of grain in sacks were piled along the tracks and in the freight yards, but under matting shelters. Near here, too, large three-mule loads of dry earth compost were going to the fields and men were busy pulverizing and mixing it on the threshing floors preparatory for use. Nearly all crops growing were one or another of the millets, but considerable areas were

yet unplanted and on these cattle, horses, mules and donkeys were feeding and eight more loads of very bright
new made hay crossed the track.

When the train reached Sinminfu where the railway
turns abruptly eastward to cross the Liao ho to reach
Mukden we saw the first extensive massing of the huge bean
cakes for export, together with enormous quantities of soy
beans in sacks piled along the railway and in the freight
yards or loaded on cars made up in trains ready to move.
Leaving this station we passed among fields of grain looking decidedly yellow, the first indication we had seen in
China of crops nitrogen-hungry and of soils markedly defi-
cient in available nitrogen. Beyond the next station the
fields were decidedly spotted and uneven as well as yellow,
recalling conditions so commonly seen at home and which
had been conspicuously absent here before. Crossing the
Liao ho with its broad channel of shifting sands, the river
carrying the largest volume of water we had yet seen,
but the stream very low and still characteristic of the close
of the dry season of semi-arid climates, we soon reached
another station where the freight yards and all of the
space along the tracks were piled high with bean cakes
and yet the fields about were reflecting the impoverished
condition of the soil through the yellow crops and their
uneven growth on the fields.

Since the Japanese–Russian war the shipments of soy
beans and of bean cake from Manchuria have increased
enormously. Up to this time there had been exports to
the southern provinces of China where the bean cakes were
used as fertilizers for the rice fields, but the new extensive markets have so raised the price that in several instances we were informed they could not then afford to
use bean cake as fertilizer. From Newchwang alone, in
1905, between January 1st and March 31st, there went
abroad 2,286,000 pounds of beans and bean cake, but in
1906 the amount had increased to 4,883,000 pounds. But a
report published in the Tientsin papers as official, while
we were there, stated that the value of the export of

bean cake and soy beans from Dalny for the months ending
March 31st had been, in 1909, only $1,635,000, gold, com-
pared with $3,065,000 in the corresponding period of 1908,
and of $5,120,000 in 1907, showing a marked decrease.

Edward C. Parker, writing from Mukden for the Review
of Reviews, stated: ''The bean cake shipments from
Newchwang, Dalny and Antung in 1908 amounted to
515,198 tons; beans, 239,298 tons; bean oil, 1930 tons;
having a total value of $15,016,649 (U. S. gold)''.

According to the composition of soy beans as indicated
in Hopkins' table of analyses, these shipments of beans
and bean cake would remove an aggregate of 6171 tons
of phosphorus, 10,097 tons of potassium, and 47,812 tons
of nitrogen from Manchurian soils as the result of export
for that year. Could such a rate have been maintained
during two thousand years there would have been sold
from these soils 20,194,000 tons of potassium; 12,342,000
tons of phosphorus and 95,624,000 tons of nitrogen; and
the phosphorus, were it thus exported, would have exceeded more than threefold all thus far produced in the
United States; it would have exceeded the world's output
in 1906 more than eighteen times, even assuming that all
phosphate rock mined was seventy-five per cent pure.

The choice of the millets and the sorghums as the staple
bread crops of northern China and Manchuria has been
quite as remarkable as the selection of rice for the more
southern latitudes, and the two togetner have played a
most important part in determining the high maintenance
efficiency of these people. In nutritive value these grains
rank well with wheat; the stems of the larger varieties
are extensively used for both fuel and building material
and the smaller forms make excellent forage and have
been used directly for maintaining the organic content of
the soil. Their rapid development and their high endurance of drought adapt them admirably to the climate of
north China and Manchuria where the rains begin only
after late June and where weather too cold for growth
comes earlier in the fall. The quick maturity of these crops

also permits them to be used to great advantage even
throughout the south, in their systems of mutiple cropping
so generally adopted, while their great resistance to
drought, being able to remain at a standstill for a long time
when the soil is too dry for growth and yet be able to push
ahead rapidly when favorable rains come, permits them
to be used on the higher lands generally where water is
not available for irrigation.

In the Shantung province the large millet, sorghum or
kaoliang, yields as high as 2000 to 3000 pounds of seed
per acre, and 5600 to 6000 pounds of air-dry stems, equal
in weight to 1.6 to 1.7 cords of dry oak wood. In the
region of Mukden, Manchuria, its average yield of seed
is placed at thirty-five bushels of sixty pounds weight per
acre, and with this comes one and a half tons of fuel or
of building material. Hosie states that the kaoliang is
the staple food of the population of Manchuria and the
principal grain food of the work animals. The grain is
first washed in cold water and then poured into a kettle
with four times its volume of boiling water and cooked
for an hour, without salt, as with rice. It is eaten with
chopsticks with boiled or salted vegetables. He states that
an ordinary servant requires about two pounds of this
grain per day, and that a workman at heavy labor will
take double the amount. A Chinese friend of his, keeping
five servants, supplied them with 240 pounds of millet
per month, together with 16 pounds of native flour,
regarded as sufficient for two days, and meat for two
days, the amount not being stated. Two of the small millets (*Setaria Italica* and *Panicum milliaceum*), wheat,
maize and buckwheat are other grains which are used as
food but chiefly to give variety and change of diet.

Very large quantities of matting and wrappings are also
made from the leaves of the large millet, which serve many
purposes corresponding with the rice mattings and bags
of Japan and southern China.

The small millets, in Shantung, yield as high as 2700
pounds of seed and 4800 pounds of straw per acre. In

Japan, in the year 1906, there were grown 737,719 acres of foxtail, barnyard and proso millet, yielding 17,084,000 bushels of seed or an average of twenty-three bushels per acre. In addition to the millets, Japan grew, the same year, 5,964,300 bushels of buckwheat on 394,523 acres, or an average of fifteen bushels per acre. The next engraving, Fig. 205, shows a crop of millet already six inches

Fig. 205.—Field of millet planted between rows of windsor beans. Chiba, Japan.

high planted between rows of windsor beans which had matured about the middle of June. The leaves had dropped, the beans had been picked from the stems, and a little later, when the roots had had time to decay the bean stems would be pulled and tied in bundles for use as fuel or for fertilizer.

We had reached Mukden thoroughly tired after a long day of continuous close observation and writing. The Astor House, where we were to stop, was three miles from the station and the only conveyance to meet the train was a four-seated springless, open, semi-baggage carryall and it was a full hour lumbering its way to our hotel. But

Fig. 206.—A Manchu lady and servant. (After Hosie.)

here as everywhere in the Orient the foreigner meets scenes and phases of life competent to divert his attention from almost any discomfort. Nothing could be more striking than the peculiar mode the Manchu ladies have of dressing their hair, seen in Fig. 206, many instances of which were passed on the streets during this early evening ride. It was fearfully and wonderfully done, laid in the smoothest, glossiest black, with nearly the lateral spread of the tail of a turkey cock and much of the backward curve of that of the rooster; far less attractive than the plainer, refined, modest, yet highly artistic style adopted by either Chinese or Japanese ladies.

The journey from Mukden to Antung required two days, the train stopping for the night at Tsaohokow. Our route lay most of the way through mountainous or steep hilly country and our train was made up of diminutive coaches drawn by a tiny engine over a three-foot two-inch narrow guage track of light rails laid by the Japanese during the war with Russia, for the purpose of moving their armies and supplies to the hotly contested fields in the Liao and Sungari plains. Many of the grades were steep, the curves sharp, and in several places it was necessary to divide the short train to enable the engines to negotiate them.

To the southward over the Liao plain the crops were almost exclusively millet and soy beans, with a little barley, wheat, and a few oats. Between Mukden and the first station across the Hun river we had passed twenty-four good sized fields of soy beans on one side of the river and twenty-two on the other, and before reaching the hilly country, after travelling a distance of possibly fifteen miles, we had passed 309 other and similar fields close along the track. In this distance also we had passed two of the monuments erected by the Japanese, marking sites of their memorable battles. These fields were everywhere flat, lying from sixteen to twenty feet above the beds of the nearly dry streams, and the cultivation was mostly being done with horses or cattle.

After leaving the plains country the railway traversed a narrow winding valley less than a mile wide, with gradient so steep that our train was divided. Fully sixty per cent of the hill slopes were cultivated nearly to the summit and yet rising apparently more than one in three to five feet, and the uncultivated slopes were closely wooded with young trees, few more than twenty to thirty feet high, but in blocks evidently of different ages. Beyond the pass many of the cultivated slopes have walled terraces. We crossed a large stream where railway ties were being rafted down the river. Just beyond this river the train was again divided to ascend a gradient of one in thirty, reaching the summit by five times switching back, and matched on the other side of the pass by a down grade of one in forty.

At many of the farm houses in the narrow valleys along the way large rectangular, flat topped compost piles were passed, thirty to forty inches high and twenty, thirty, forty and even in one case as much as sixty feet square on the ground. More and more it became evident that these mountain and hill lands were originally heavily wooded and that the new growth springs up quickly, developing rapidly. It was clear also that the custom of cutting over these wooded areas at frequent intervals is very old, not always in the same stage of growth but usually when the trees are quite small. Considerable quantities of cordwood were piled at the stations along the railway and were being loaded on the cars. This was always either round wood or sticks split but once; and much charcoal, made mostly from round wood or sticks split but once, was being shipped in sacks shaped like those used for rice, seen in Fig. 180. Some strips of the forest growth had been allowed to stand undisturbed apparently for twenty or more years, but most areas have been cut at more frequent intervals, often apparently once in three to five, or perhaps ten, years.

At several places on the rapid streams crossed, prototypes of the modern turbine water-wheel were installed, doing duty grinding beans or grain. As with native machinery everywhere in China, these wheels were reduced to the lowest terms and the principle put to work almost unclothed. These turbines were of the downward discharge type, much resembling our modern windmills, ten to sixteen feet in diameter, set horizontally on a vertical axis rising through the floor of the mill, with the vanes surrounded by a rim, the water dropping through the wheel, reacting when reflected from the obliquely set vanes. American engineers and mechanics would pronounce these very crude, primitive and inefficient. A truer view would regard them as examples of a masterful grasp of principle

Fig. 207.—Gathering of Koreans in holiday attire, on their national "Swing day."

by some man who long ago saw the unused energy of the stream and succeeded thus in turning it to account.

Both days of our journey had been bright and very warm and, although we took the train early in the morning at Mukden, a young Japanese anticipated the heat, entering the train clad only in his kimono and sandals, carrying a suitcase and another bundle. He rode all day, the most comfortably, if immodestly, clad man on the train, and the next morning took his seat in front of us clad in the same garb, but before the train reached Antung he took down his suitcase and then and there, deliberately attired himself in a good foreign suit, folding his kimono and packing it away with his sandals.

From Antung we crossed the Yalu on the ferry to New Wiju at 6:30 A. M., June 22, and were then in quite a different country and among a very different people, although all of the railway officials, employes, police and guards were Japanese, as they had been from Mukden. At Antung and New Wiju the Yalu is a very broad slow

Fig. 208.—Group of Koreans at Gyoha, being addressed by a public speaker on Swing day.

stream resembling an arm of the sea more than a river, reminding one of the St. Johns at Jacksonville, Florida.

June 22nd proved to be one of the national festival days in Korea, called "Swing day", and throughout our entire ride to Seoul the fields were nearly all deserted and throngs of people, arrayed in gala dress, appeared all along the line of the railway, sometimes congregating in bodies of two to three thousand or more, as seen in Fig. 207. Many swings had been hung and were being enjoyed by the young people. Boys and men were bathing in all sorts of "swimming holes" and places. So too, there were many large open air gatherings being addressed by public speakers, one of which is seen in Fig. 208.

Nearly everyone was dressed in white outer garments made from some fabric which although not mosquito netting was nearly as open and possessed of a remarkable stiffness which seemed to take and retain every dent with

Fig. 209.—Group of five Korean women in their stiff white clothing.

astonishing effect and which was sufficiently transparent to reveal a third undergarment. The full out-standing skirts of five Korean women may be seen in Fig. 209, and the trousers which went with these were proportionately full but tied close about the ankles. The garments seemed to be possessed of a powerful repulsion which held them quite apart and away from the person, no doubt contributing much to comfort. It was windy but one of those hot sultry, sticky days, and it made one feel cool to see these open garments surging in the wind.

The Korean men, like the Chinese, wear the hair long but not braided in a queue. No part of the head is shaved but the hair is wound in a tight coil on the top of the head, secured by a pin which, in the case of the Korean who rode in our coach from Mukden to Antung, was a modern, substantial ten-penny wire nail. The tall, narrow, conical crowns of the open hats, woven from thin bamboo

Fig. 210.—Group of Korean farm houses with thatched roofs and earthern walls, standing at the foot of wooded hills.

splints, are evidently designed to accommodate this style of hair dressing as well as to be cool.

Here, too, as in China and Manchuria, nearly all crops are planted in rows, including the cereals, such as wheat, rye, barley and oats. We traversed first a flat marshy country with sandy soil and water not more than four feet below the surface where, on the lowest areas a close ally of our wild flower-de-luce was in bloom. Wheat was coming into head but corn and millet were smaller than in Manchuria. We had left New Wiju at 7:30 in the morning and at 8:15 we passed from the low land into a hill country with narrow valleys. Scattering young pine, seldom more than ten to twenty-five feet high, occupied the slopes and as we came nearer the hills were seen to be clothed with many small oak, the sprouts clearly not more than one or two years old. Roofs of dwellings in the country were usually thatched with straw laid after the manner of shingles, as may be seen in Fig. 210, where the hills beyond show the low tree growth referred to, but here unusually dense. Bundles of pine boughs, stacked and sheltered from the weather, were common along the way and evidently used for fuel.

At 8:25 we passed through the first tunnel and there were many along the route, the longest requiring thirty seconds for the passing of the train. The valley beyond was occupied by fields of wheat where beans were planted between the rows. Thus far none of the fields had been as thoroughly tilled and well cared for as those seen in China, nor were the crops as good. Further along we passed hills where the pines were all of two ages, one set about thirty feet high and the others twelve to fifteen feet or less, and among these were numerous oak sprouts. Quite possibly these are used as food for the wild silkworms. In some places appearances indicate that the oak and other deciduous growth, with the grass, may be cut annually and only the pines allowed to stand for longer periods. As we proceeded southward and had passed Kosui the young oak sprouts were seen to cover the hills, often stretching over the slopes much like a regular crop, standing at a hight of two to four feet, and fresh bundles of these sprouts were seen at houses along the foot of the slopes, again suggesting that the leaves may be for the tussur silkworms although the time appears late for the first moulting. After we had left Seoul, entering the broader valleys where rice was more extensively grown, the using of the oak boughs and green grass brought down from the hill lands for green manure became very extensive.

After the winter and early spring crops have been harvested the narrow ridges on which they are grown are turned into the furrows by means of their simple plow drawn by a heavy bullock, different from the cattle in China but closely similar to those in Japan. The fields are then flooded until they have the appearance seen in Fig. 12. Over these flooded ridges the green grass and oak boughs are spread, when the fields are again plowed and the material worked into the wet soil. If this working is not

Fig. 211.—General view across valley, showing Korean rice fields being transplanted, and in the foreground fertilized with green herbage from the hill lands.

completely successful men enter the fields and tramp the surface until every twig and blade is submerged. The middle section in this illustration has been fitted and transplanted; in front of it and on the left are two other fields once plowed but not fertilized; those far to the right have had the green manure applied and the ground plowed a second time but not finished, and in the immediate foreground the grass and boughs have been scattered but the second plowing is not yet done.

We passed men and bullocks coming from the hill lands loaded with this green herbage and as we proceeded to-

wards Fusan more and more of the hill area was being
made to contribute materials for green manure for the cultivated fields.   The foreground of Fig. 211 had been thus
treated and so had the field in Fig. 212, where the man was
engaged in tramping the dressing beneath the surface.   In
very many cases this material was laid along the margin of
the paddies; in other cases it had been taken upon the fields
as soon as the grain was cut and was lying in piles among
the bundles; while in still other cases the material for

Fig. 212.—Rice paddy covered with oak leaves and grass brought down from
the hills, one half of which has been tramped beneath the surface by the
laborer at work.

green manure had been carried between the rows while the
grain was still standing, but nearly ready to harvest.   In
some fields a full third of a bushel of the green stuff had
been laid down at intervals of three feet over the whole area.
In other cases piles of ashes alternated with those of herbage, and again manure and ashes mixed had been distributed in alternate piles with the green manure.

In still other cases we saw untreated straw distributed
through the fields awaiting application.   At Shindo this
straw had the appearance of having been dipped in or
smeared with some mixture, apparently of mud and ashes
or possibly of some compost which had been worked into
a thin paste with water.

After passing Keizan, mountain herbage had been brought down from the hills in large bales on cleverly constructed racks saddled to the backs of bullocks, and in one field we saw a man who had just come to his little field with an enormous load borne upon his easel-like packing appliance. Thus we find the Koreans also adopting the rice crop, which yields heavily under conditions of abun-

Fig. 213.—Rice paddies at head of mountain valley, with scattering pines in the hill lands beyond.

dant water; we find them supplementing a heavy summer rainfall with water from their hills, and bringing to their fields besides both green herbage for humus and organic matter, and ashes derived from the fuel coming also from the hills, in these ways making good the unavoidable losses through intense cropping.

The amount of forest growth in Korea, as we saw it, in proximity to the cultivated valleys, is nowhere large and is fairly represented in Figs. 210, 213 and 214. There were clear evidences of periodic cutting and considerable amounts of cordwood split from timber a foot through were being brought to the stations on the backs of cattle. In some places there was evident and occasionally very serious soil erosion, as may be seen in Fig. 214, one such region being passed just before reaching Kinusan, but generally the hills are well rounded and covered with a low growth of shrubs and herbaceous plants.

Southernmost Korea has the latitude of the northern boundary of South Carolina, Georgia, Alabama and Mississippi, while the northeast corner attains that of Madison,

Fig. 214.—Looking across fields of wheat at an eroding hillside over which forest growth is being allowed to spread.

Wisconsin, and the northern boundary of Nebraska, the country thus spanning some nine degrees and six hundred miles of latitude. It has an area of some 82,000 square miles, about equaling the state of Minnesota, but much of its surface is occupied by steep hill and mountain land. The rainy season had not yet set in, June 23rd. Wheat and the small grains were practically all harvested southward of Seoul and the people were everywhere busy with their flails threshing in the open, about the dwellings or in the fields, four flails often beating together on the same lot of grain. As we journeyed southward the valleys and the fields became wider and more extensive, and the crops, as well as the cultural methods, were clearly much better.

Neither the foot-power, animal-power, nor the wooden chain pump of the Chinese were observed in Korea in use for lifting water, but we saw many instances of the long handled, spoonlike swinging scoop hung over the water by a cord from tall tripods, after the manner seen in Fig. 215, each operated by one man and apparently with high efficiency for low lifts. Two instances also were observed of

Fig. 215.—Korean swinging scoop for irrigation where the water is raised three or four feet.

the form of lift seen in Fig. 173, where the man walks the circumference of the wheel, so commonly observed in Japan. Much hemp was being grown in southern Korea but everywhere on very small isolated areas which flecked the landscape with the deepest green, each little field probably representing the crop of a single family.

It was 6:30 P. M. when our train reached Fusan after a hot and dusty ride. The service had been good and fairly comfortable but the ice-water tanks of American trains were absent, their place being supplied by cooled bottled waters of various brands, including soda-water, sold by Japanese boys at nearly every important station. Close connection was made by trains with steamers to and from Japan and we went directly on board the Iki Maru which was to weigh anchor for Moji and Shimonoseki at 8 P. M. Although small, the steamer was well equipped, providing the best of service. We were fortunate in having a smooth passage, anchoring at 6:30 the next morning and making close connection with the train for Nagasaki, landing at the wharf with the aid of a steam launch.

Our ride by train through the island of Kyushu carried us through scenes not widely different from those we had just left. The journey was continuously among fields of rice, with Korean features strongly marked but usually under better and more intensified culture, and the season, too, was a little more advanced. Here the plowing was being done mostly with horses instead of the heavy bullocks so exclusively employed in Korea. Coming from China into Korea, and from there into Japan, it appeared very clear that in agricultural methods and appliances the Koreans and Japanese are more closely similar than the Chinese and Koreans, and the more we came to see of the Japanese methods the more strongly the impression became fixed that the Japanese had derived their methods either from the Koreans or the Koreans had taken theirs more largely from Japan than from China.

It was on this ride from Moji to Nagasaki that we were introduced to the attractive and very satisfactory manner of serving lunches to travelers on the trains in Japan. At important stations hot tea is brought to the car windows in small glazed, earthernware teapots provided with cover and bail, and accompanied with a teacup of the same ware. The set and contents could be purchased for five sen, two and a half cents, our currency. All tea is served without milk or sugar. The lunches were very substantial and put together in a neat sanitary manner in a three-compartment wooden box, carefully made from clear lumber joined with wooden pegs and perfect joints. Packed in the cover we found a paper napkin, toothpicks and a pair of chopsticks. In the second compartment there were thin slices of meat, chicken and fish, together with bamboo sprouts, pickles, cakes and small bits of salted vegetables, while the lower and chief compartment was filled with rice cooked quite stiff and without salt, as is the custom in the three countries. The box was about six inches long, four inches deep and three and a half inches wide. These lunches are handed to travelers neatly wrapped in spotless thin white paper daintily tied with a bit of color, all in exchange for 25 sen,—12.5 cents. Thus for fifteen cents the traveler is handed, through the car window, in a respectful manner, a square meal which he may eat at his leisure.

XVII.

RETURN TO JAPAN.

We had returned to Japan in the midst of the first rainy season, and all the day through, June 25th, and two nights, a gentle rain fell at Nagasaki, almost without interruption. Across the narrow street from Hotel Japan were two of its guest houses, standing near the front of a wall-faced terrace rising twenty-eight feet above the street and facing the beautiful harbor. They were accessible only by winding stone steps shifting on paved landings to continue the ascent between retaining walls overhung with a wealth of shrubbery clothed in the densest foliage, so green and liquid in the drip of the rain, that one almost felt like walking edgewise amid stairs lest the drip should leave a stain. Over such another series of steps, but longer and more winding, we found our way to the American Consulate where in the beautifully secluded quarters Consul-General Scidmore escaped many annoyances of settling the imagined petty grievances arising between American tourists and the ricksha boys.

Through the kind offices of the Imperial University of Sapporo and of the National Department of Agriculture and Commerce, Professor Tokito met us at Nagasaki, to act as escort through most of the journey in Japan. Our first visit was to the prefectural Agricultural Experiment Station at Nagasaki. There are forty others in the four main islands, one to an average area of 4280 square miles, and to each 1,200,000 people.

The island of Kyushu, whose latitude is that of middle Mississippi and north Louisiana, has two rice harvests, and gardeners at Nagasaki grow three crops, each year. The gardener and his family work about five tan, or a little less than one and one-quarter acres, realizing an annual return of some $250 per acre. To maintain these earnings fertilizers are applied rated worth $60 per acre, divided between the three crops, the materials used being largely the wastes of the city, animal manure, mud from the drains, fuel ashes and sod, all composted together. If this expenditure for fertilizers appears high it must be remembered that nearly the whole product is sold and that there are three crops each year. Such intense culture requires a heavy return if large yields are maintained. Good agricultural lands were here valued at 300 yen per tan, approximately $600 per acre.

When returning toward Moji to visit the Agricultural Experiment Station of Fukuoka prefecture, the rice along the first portion of the route was standing about eight inches above the water. Large lotus ponds along the way occupied areas not readily drained, and the fringing fields between the rice paddies and the untilled hill lands were bearing squash, maize, beans and Irish potatoes. Many small areas had been set to sweet potatoes on close narrow ridges, the tops of which were thinly strewn with green grass, or sometimes with straw or other litter, for shade and to prevent the soil from washing and baking in the hot sun after rains. At Kitsu we passed near Government salt works, for the manufacture of salt by the evaporation of sea water, this industry in Japan, as in China, being a Government monopoly.

Many bundles of grass and other green herbage were collected along the way, gathered for use in the rice fields. In other cases the green manure had already been spread over the flooded paddies and was being worked beneath the surface, as seen in Fig. 216. At this time the hill lands were clothed in the richest, deepest green but the tree growth was nowhere large except immediately about temples, and was usually in distinct small areas with sharp boundaries occasioned by differences in age. Some tracts had been very recently cut; others were in their second, third or fourth years; while others still carried a growth of perhaps seven to ten years. At one village many bundles of the brush fuel had been gathered from an adjacent area, recently cleared.

A few fields were still bearing their crop of soy beans planted in February between rows of grain, and the green herbage was being worked into the flooded soil, for the crop of rice. Much compost, brought to the fields, was stacked with layers of straw between, laid straight, the alternate courses at right angles, holding the piles in rectangular form with vertical sides, some of which were four to six feet high and the layers of compost about six inches thick.

Just before reaching Tanjiro a region is passed where orchards of the candleberry tree occupy high leveled areas between rice paddies, after the manner described for the mulberry orchards in Chekiang, China. These trees, when seen from a distance, have quite the appearance of our apple orchards.

At the Fukuoka Experiment Station we learned that the usual depth of plowing for the rice fields is three and a half to four and a half inches, but that deeper plowing gives somewhat larger yields. As an average of five years trials, a depth of seven to eight inches increased the yield from seven to ten per cent over that of the usual depth. In this prefecture grass from the bordering hill lands is applied to the rice fields at rates ranging from 3300 to 16,520 pounds green weight per acre, and, according to analyses given, these amounts would carry to the fields from 18 to 90 pounds of nitrogen; 12.4 to 63.2 pounds of potassium, and 2.1 to 10.6 pounds of phosphorus per acre.

Where bean cake is used as a fertilizer the applications may be at the rate of 496 pounds per acre, carrying 33.7 pounds of nitrogen, nearly 5 pounds of phosphorus and 7.4 pounds of potassium. The earth composts are chiefly applied to the dry land fields and then only after they are well rotted, the fermentation being carried through at least sixty days, during which the material is turned three times for aeration, the work being done at the home. When used on the rice fields where water is abundant the composts are applied in a less fermented condition.

The best yields of rice in this prefecture are some eighty bushels per acre, and crops of barley may even exceed this, the two crops being grown the same year, the rice following the barley. In most parts of Japan the grain food of the laboring people is about 70 per cent naked barley mixed with 30 per cent of rice, both cooked and used in the same manner. The barley has a lower market value and its use permits a larger share of the rice to be sold as a money crop.

The soils are fertilized for each crop every year and the prescription for barley and rice recommended by the Experiment Station, for growers in this prefecture, is indicated by the following table:

FERTILIZATION FOR NAKED BARLEY.

Fertilizers.		Pounds per acre.		
		N	P	K
Manure compost	6,613	33.0	7.4	33.8
Rape seed cake	330	16.7	2.8	3.5
Night soil	4,630	26.4	2.6	10.2
Superphosphate	132	....	9.9	....
Sum	11,705	76.1	22.7	47.5

FERTILIZATION FOR PADDY RICE.

Manure compost	5,291	26.4	5.9	27.1
Green manure, soy beans	3,306	19.2	1.1	19.6
Soy bean cake	397	27.8	1.7	6.4
Superphosphate	198	....	12.8	....
Sum	9,192	73.4	21.5	53.1
Total for year	20,897	149.5	44.2	100.6

Where these recommendations are followed there is an annual application of fertilizer material which aggregates some ten tons per acre, carrying about 150 pounds of nitrogen, 44 pounds of phosphorus and 100 pounds of potassium. The crop yields which have been associated with these applications on the Station fields are about forty-nine bushels of barley and fifty bushels of rice per acre.

The general rotation recommended for this portion of Japan covers five years and consists of a crop of wheat or naked barley the first two years with rice as the summer crop; in the third year *genge,* "pink clover" (*Astragalus sinicus*) or some other legume for green manure is the winter crop, rice following in the summer; the fourth year rape is the winter crop, from which the seed is saved and the ash of the stems returned to the soil, or rarely the

Fig. 216.—Working green herbage into a flooded rice paddy for green manure, preparatory for the following crop of rice.

stems themselves may be turned under; on the fifth and last year of the rotation the broad kidney or windsor bean is the winter crop, preceding the summer crop of rice. This rotation is not general yet in the practice of the farmers of the section, they choosing rape or barley and in February plant windsor or soy beans between the rows for green manure to use when the rice comes on.

It was evident from our observations that the use of composts in fertilizing was very much more general and extensive in China than it was in either Korea or Japan, but, to encourage the production and use of compost fertilizers, this and other prefectures have provided subsidies which permit the payment of $2.50 annually to those farmers who prepare and use on their land a compost heap covering twenty to forty square yards, in accordance with specified directions given.

The agricultural college at Fukuoka was not in session the day of our visit, it being a holiday usually following the close of the last transplanting season. One of the main buildings of the station and college is seen in Fig. 217, and Figs. 218, 219 and 220, placed together from left to right in the order of their numbers, form a panoramic view

Fig. 217.—One of the main buildings of the Fukuoka Experiment Station.

of the station grounds and buildings with something of the beautiful landscape setting. There is nowhere in Japan the lavish expenditure of money on elaborate and imposing architecture which characterizes American colleges and stations, but in equipment for research work, both as to professional staff and appliances, they compare favorably with similar institutions in America. The dormitory system was in vogue in the college, providing room and board at eight yen per month or four dollars of our currency. Eight students were assigned to one commodious room, each provided with a study table, but beds were mattresses spread upon the matting floor at night and compactly stored on closet shelves during the day.

The Japanese plow, which is very similar to the Korean

Fig. 218.—View of station grounds and buildings, Fukuoka Experiment Station, **Japan.**

Fig. 219.—View of station grounds and buildings, Fukuoka Experiment Station, Japan.

Fig. 220.—View of station grounds and buildings, Fukuoka Experiment Station, Japan.

type, may be seen in Fig. 221, the one on the right costing 2.5 yen and the other 2 yen. With the aid of the single handle and the sliding rod held in the right hand, the course of the plow is directed and the plow tilted in either direction, throwing the soil to the right or the left.

The nursery beds for rice breeding experiments and variety tests by this station are shown in Fig. 222. Although these plots are flooded the marginal plants, adjacent to the free water paths, were materially larger than those within and had a much deeper green color, showing better feeding, but what seemed most strange was the fact that these stronger plants are never used in transplanting, as they do not thrive as well as those less vigorous.

We left the island of Kyushu in the evening of June 29th, crossing to the main island of Honshu, waiting in Shimonoseki for the morning train. The rice planted valleys near Shimonoseki were relatively broad and the paddies had all been recently set in close rows about a foot apart and in hills in the rows. Mountain and hill lands were closely wooded, largely with coniferous trees about the base but toward and at the summits, especially on the south slopes, they were green only with herbage cut for fertilizing and feeding stock. Many very small trees, often not more than one foot high, were growing on the recently cut-over areas; tall slender graceful bamboos clustered along the way and everywhere threw wonderful beauty into the landscape. Cartloads of their slender stems, two to four inches in diameter at the base and twenty or more feet long, were moving along the generally excellent, narrow, seldom fenced roads, such as seen in Fig. 223. On the borders and pathways between rice paddies many small stacks of straw were in waiting to be laid between the rows of transplanted rice, tramped beneath the water and overspread with mud to enrich the soil. The farmers here, as elsewhere, must contend against the scouring rush, varieties of grass and our common pigweeds, even in the rice fields. The large area of moun-

25

tain and hill land compared with that which could be
tilled, and the relatively small area of cultivated land
not at this time under water and planted to rice persisted
throughout the journey.

Fig. 221.—Two Japanese plows.

If there could be any monotony for the traveller new
to this land of beauty it must result from the quick shifting of scenes and in the way the landscapes are pieced

together, out-doing the craziest patchwork woman ever attempted; the bits are almost never large; they are of every shape, even puckered and crumpled and tilted at all angles. Here is a bit of the journey: Beyond Habu the foothills are thickly wooded, largely with conifers. The valley is extremely narrow with only small areas for rice. Bamboo are growing in congenial places and we pass bundles of wood cut to stove length, as seen in Fig. 224. Then we cross a long narrow valley practically all in rice, and then another not half a mile wide, just before reach-

Fig. 222.—Plant breeding and variety test nursery rice plats, at Fukuoka Experiment Station.

ing Asa. Beyond here the fields become limited in area with the bordering low hills recently cut over and a new growth springing up over them in the form of small shrubs among which are many pine. Now we are in a narrow valley between small rice fields or with none at all, but dash into one more nearly level with wide areas in rice chiefly on one side of the track just before reaching Onoda at 10:30 A. M. and continuing three minutes ride beyond, when we are again between hills without fields and where the trees are pine with clumps of bamboo. In four minutes more we are among small rice paddies and at 10:35 have passed another gap and are crossing another valley checkered with rice fields and lotus ponds, but in one minute more the hills have closed in, leaving only room for the track. At 10:37 we are running along a narrow valley with its terraced rice paddies where many of the hills show naked soil among the bamboo, scattering pine and other small trees; then we are out among garden patches thickly mulched with straw. At 10:38 we are between higher hills with but narrow areas for rice stretching close

Fig. 223.—Public highway in Japan.

along the track, but in two minutes these are passed and we are among low hills with terraced dry fields. At 10:42 we are spinning along the level valley with its rice, but are quickly out again among hills with naked soil where erosion was marked. This is just before passing Funkai where we are following the course of a stream some sixty feet wide with but little cultivated land in small areas. At 10:47 we are again passing narrow rice fields near the track where the people are busy weeding with their hands, half knee-deep in water. At 10:53 we enter a broader valley stretching far to the south and seaward, but we had crossed it in one minute, shot through another gap, and at 10:55 are traversing a much broader valley largely given over to rice, but where some of the paddies were bearing matting rush set in rows and in hills after the manner of rice. It is here we pass Oyou and just beyond cross a stream confined between levees built some distance back from either bank. At 11:17 this plain is left and we enter a narrow valley without fields. Thus do most of the agricultural

Fig. 224.—Transporting wood to market. Japan.

lands of Japan lie in the narrowest valleys, often steeply sloping, and into which jutting spurs create the greatest irregularity of boundary and slope.

The journey of this day covered 350 miles in fourteen hours, all of the way through a country of remarkable and peculiar beauty which can be duplicated nowhere outside the mountainous, rice-growing Orient and there only during fifteen days closing the transplanting season. There were neither high mountains nor broad valleys, no great rivers and but few lakes; neither rugged naked rocks, tall

Fig. 225.—Looking up a terraced valley between Hongo and Fukuyama, Japan.

forest trees nor wide level fields reaching away to unbroken horizons. But the low, rounded, soil-mantled mountain tops clothed in herbaceous and young forest growth
fell everywhere into lower hills and these into narrow
steep valleys which dropped by a series of water-level
benches, as seen in Fig. 225, to the main river courses.
Each one of these millions of terraces, set about by its
raised rim, was a silvery sheet of water dotted in the daint-

Fig. 226.—Group of houses standing in rice paddies, on edge of terraces, surrounded by water.

iest manner with bunches of rice just transplanted, but
not so close nor yet so high and over-spreading as to ob-
scure the water, yet quite enough to impart to the surface
a most delicate sheen of green; and the grass-grown narrow rims retaining the water in the basins, cemented them
into series of the most superb mosaics, shaped into the valley bottoms by artizan artists perhaps two thousand years
before and maintained by their descendants through all
the years since, that on them the rains and fertility from
the mountains and the sunshine from heaven might be
transformed by the rice plant into food for the families

and support for the nation. Two weeks earlier the aspect of these landscapes was very different, and two weeks later the reflecting water would lie hidden beneath the growing and rapidly developing mantle of green, to go on changing until autumn, when all would be overspread with the ripened harvest of grain. And what intensified the beauty of it all was the fact that only along the widest valley bottoms were the mosaics level, except the water surface of each individual unit and these were always small. At one time we were riding along a descending series of steps and then along another rising through a winding valley to disappear around a projecting spur, and anywhere in the midst of it all might be standing Japanese cottages or villas with the water and the growing rice literally almost against the walls, as seen in Fig. 226, while a near-by high terrace might hold its water on a level with the chimneytops. Can one wonder that the Japanese loves his country or that they are born and bred landscape artists?

Just before reaching Hongo there were considerable areas thrown into long narrow, much raised, east and west beds under covers of straw matting inclined at a slight angle toward the south, some two feet above the ground but open toward the north. What crop may have been grown here we did not learn but the matting was apparently intended for shade, as it was hot midsummer weather, and we suspect it may have been ginseng. It was here, too, that we came into the region of the culture of matting rush, extensively grown in Hiroshima and Okayama prefectures, but less extensively all over the empire. As with rice, the rush is first grown in nursery beds from which it is transplanted to the paddies, one acre of nursery supplying sufficient stock for ten acres of field. The plants are set twenty to thirty stalks in a hill in rows seven inches apart with the hills six inches from center to center in the row. Very high fertilization is practiced, costing from 120 to 240 yen per acre, or $60 to $120 annually, the fertilizer consisting of bean cake and plant ashes, or in recent years, sometimes of sulphate of ammonia for nitrogen, and superphosphate of lime. About ten per cent of the amount of fertilizer required for the crop is applied at the time of fitting the ground, the balance being administered from time to time as the season advances. Two crops of the rush may be taken from the same ground each year or it is grown in rotation with rice, but most extensively on the lands less readily drained and not so

Fig. 227.—Fields of matting rush with recently transplanted rice, and Government salt fields in the background.

well suited for other crops. Fields of the rush, growing in alternation with rice, are seen in Fig. 45, and in Fig. 227, with the Government salt fields lying along the seashore beyond.

With the most vigorous growth the rush attain a hight exceeding three feet and the market price varies materially with the length of the stems. Good yields, under the best culture, may be as high as 6.5 tons per acre of the dry stems but the average yield is less, that of 1905 being 8531

Fig. 228.—Group of Japanese girls playing the game of flower cards, in the usual attitude of sitting on the matting-covered floor.

Fig. 229.—Interior view of a well furnished guest room in a Japanese inn, where the meals are served on the matting floor and the bed is laid.

pounds, for 9655 acres. The value of the product ranges from \$120 to \$200 per acre.

It is from this material that mats are woven in standard sizes, to be laid over padding, upholstering the floors which are the seats of all classes in Japan, used in the manner seen in Fig. 228 and in Fig. 229, which is a completely furnished guest room in a first class Japanese inn, finished in natural unvarnished wood, with walls of sliding panels of translucent paper, which may open upon a porch, into a hallway or into another apartment; and with its bouquet, which may consist of a single large shapely branch of the purple leaved maple, having the cut end charred to preserve it fresh for a longer time, standing in water in the vase.

"Two little maids I've heard of, each with a pretty taste,
Who had two little rooms to fix and not an hour to waste.
Eight thousand miles apart they lived, yet on the selfsame day
The one in Nikko's narrow streets, the other on Broadway,
They started out, each happy maid her heart's desire to find,
And her own dear room to furnish just according to her mind.

When Alice went a-shopping, she bought a bed of brass,
A bureau and some chairs and things and such a lovely glass
To reflect her little figure—with two candle brackets near—
And a little dressing table that she said was simply dear!
A book shelf low to hold her books, a little china rack,
And then, of course, a bureau set and lots of bric-a-brac;
A dainty little escritoire, with fixings all her own
And just for her convenience, too, a little telephone.
Some oriental rugs she got, and curtains of madras,
With 'cunning' ones of lace inside, to go against the glass;
And then a couch, a lovely one, with cushions soft to crush,
And forty pillows, more or less, of linen, silk and plush;
Of all the ornaments besides I couldn't tell the half,
But wherever there was nothing else, she stuck a photograph.
And then, when all was finished, she sighed a little sigh,
And looked about with just a shade of sadness in her eye:
'For it needs a statuette or so—a fern—a silver stork—
Oh, something, just to fill it up!' said Alice of New York.

When little Oumi of Japan went shopping, pitapat,
She bought a fan of paper and a little sleeping mat;
She set beside the window a lily in a vase,
And looked about with more than doubt upon her pretty face:
'For, really—don't you think so?—with the lily and the fan,
It's a little overcrowded!' said Oumi of Japan."
(Margaret Johnson in St. Nicholas Magazine)

In the rural homes of Japan during 1906 there were woven 14,497,058 sheets of these floor mats and 6,628,772 sheets of other matting, having a combined value of $2,815,040, and in addition, from the best quality of rush grown upon the same ground, aggregating 7657 acres that year, there were manufactured for the export trade, fancy mattings having the value of $2,274,131. Here is a total value, for the product of the soil and for the labor put into the manufacture, amounting to $664 per acre for the area named.

At the Akashi agricultural experiment station, under the Directorship of Professor Ono, we saw some of the methods of fruit culture as practiced in Japan. He was conducting experiments with the object of improving methods of heading and training pear trees, to which reference was made on page 22. A study was also being made of the advantages and disadvantages associated with covering the fruit with paper bags, examples of which are seen in Figs. 6 and 7. The bags were being made at the time of our visit, from old newspapers cut, folded and pasted by women. Naked cultivation was practiced in the orchard, and fertilizers consisting of fish guano and superphosphate of lime were being applied twice each year in amounts aggregating a cost of twenty-four dollars per acre.

Pear orchards of native varieties, in good bearing, yield returns of 150 yen per tan, and those of European varieties, 200 yen per tan, which is at the rate of $300 and $400 per acre. The bibo so extensively grown in China was being cultivated here also and was yielding about $320 per acre.

It was here that we first met the cultivation of a variety of burdock grown from the seed, three crops being taken each season where the climate is favorable, or as one of three in the multiple crop system. It is grown for the root, yielding a crop valued at $40 to $50 per acre. One crop, planted in March, was being harvested July 1st.

During our ride to Akashi on the early morning train we passed long processions of carts drawn by cattle, horses or by men, moving along the country road which paralleled the railway, all loaded with the waste of the city of Kobe, going to its destination in the fields, some of it a distance of twelve miles, where it was sold at from 54 cents to $1.63 per ton.

At several places along our route from Shimonoseki to Osaka we had observed the application of slacked lime to the water of the rice fields, but in this prefecture, Hyogo, where the station is located, its use was prohibited in 1901, except under the direction of the station authorities, where the soil was acid or where it was needed on account of insect troubles. Up to this time it had been the custom of farmers to apply slacked lime at the rate of three to five tons per acre, paying for it $4.84 per ton. The first restrictive legislation permitted the use of 82 pounds of lime with each 827 pounds of organic manure, but as the farmers persisted in using much larger quantities, complete prohibition was resorted to.

Reference has been made to subsidies encouraging the use of composts, and in this prefecture prizes are awarded for the best compost heaps in each county, examinations being made by a committee. The composts receiving the four highest awards in each county are allowed to compete with those in other counties for a prefectural prize awarded by another committee.

The "pink clover" grown in Hyogo after rice, as a green manure crop, yields under favorable conditions twenty tons of the green product per acre, and is usually applied to about three times the area upon which it grew, at the rate of 6.6 tons per acre, the stubble and roots serving for the ground upon which the crop grew.

On July 3rd we left Osaka, going south through Sakai to Wakayama, thence east and north to the Nara Experiment Station. After passing the first two stations the route lay through a very flat, highly cultivated garden section with cucumbers trained on trellises, many squash in full bloom, with fields of taro, ginger and many other vegetables. Beyond Hamadera considerable areas of flat

Fig. 230.—Distribution of old stubble and the working of it beneath the water and mud to serve as fertilizer.

sandy land had been set close with pine, but with intervening areas in rice, where the growers were using the revolving weeder seen in Fig. 14. At Otsu broad areas are in rice but here worked with the short handled claw weeders, and stubble from a former crop had been drawn together into small piles, seen in Fig. 230, which later would be carefully distributed and worked beneath the mud.

Much of the mountain lands in this region, growing pine, is owned by private parties and the growth is cut at intervals of ten, twenty or twenty-five years, being sold on the ground to those who will come and cut it at a price of forty sen for a one-horse load, as already described, page 159.

The course from here was up the rather rapidly rising

Fig. 231.—Irrigating with the Japanese circumferential foot-power water wheel near Hashimoto, Japan.

Kiigawa valley where much water was being applied to the rice fields by various methods of pumping, among them numerous current wheels; an occasional power-pump driven by cattle; and very commonly the foot-power wheel where the man walks on the circumference, steadying himself with a long pole, as seen in the field, Fig. 231. It was here that a considerable section of the hill slope had been very recently cut over, the area showing light in the engraving. It was in the vicinity of Hashimoto on this route, too, that the two beautiful views reproduced in Figs. 151 and 152 were taken.

At the experiment station it was learned that within the prefecture of Nara, having a population of 558,314, and 107,574 acres of cultivated land, two-thirds of this was in paddy rice. Within the province there are also about one thousand irrigation reservoirs with an average depth of eight feet. The rice fields receive 16.32 inches of irrigation water in addition to the rain.

Of the uncultivated hill lands, some 2500 acres contribute green manure for fertilization of fields. Reference has been made to the production of compost for fertilizers on page 211. The amount recommended in this prefecture as a yearly application for two crops grown is:

Organic matter	3,711 to 4,640 lbs.	per acre
Nitrogen	105 to 131 lbs.	per acre
Phosphorus	35 to 44 lbs.	per acre
Potassium	56 to 70 lbs.	per acre

These amounts, on the basis of the table, p. 214, are nearly sufficient for a crop of thirty bushels of wheat, followed by one of thirty bushels of rice, the phosphorus being in excess and the potassium not quite enough, supposing none to be derived from other sources.

At the Nara hotel, one of the beautiful Japanese inns where we stopped, our room opened upon a second story veranda from which one looked down upon a beautiful, tiny lakelet, some twenty by eighty feet, within a diminutive park scarcely more than one hundred by two hundred feet, and the lakelet had its grassy, rocky banks over-hung with trees and shrubs planted in all the wild disorder and beauty of nature; bamboo, willow, fir, pine, cedar, red-leaved maple, catalpa, with other kinds, and through these, along the shore, wound a woodsy, well trodden, narrow footpath leading from the inn to a half hidden cottage apparently quarters for the maids, as they were frequently passing to and fro. A suggestion of how such wild beauty is brought right to the very doors in Japan may be gained

from Fig. 232, which is an instance of parking effect on a
still smaller scale than that described.

On the morning of July 6th, with two men for each of
our rickshas, we left the Yaami hotel for the Kyoto Experiment station, some two miles to the southwest of the
city limits. As soon as we had entered upon the country
road we found ourselves in a procession of cart men each

Fig. 232.—Beauty at home in Japan.

drawing a load of six large covered receptacles of about
ten gallons capacity, and filled with the city's waste. Before reaching the station we had passed fifty-two of these
loads, and on our return the procession was still moving
in the same direction and we passed sixty-one others, so
that during at least five hours there had moved over this
section of road leading into the country, away from the city,
not less than ninety tons of waste; along other roadways
similar loads were moving. These freight carts and those
drawn by horses and bullocks were all provided with long

racks similar to that illustrated in Fig. 108, page 197, and when the load is not sufficient to cover the full length it is always divided equally and placed near each end, thus taking advantage of the elasticity of the body to give the effect of springs, lessening the draft and the wear and tear.

One of the most common commodities coming into the city along the country roads was fuel from the hill lands, in split sticks tied in bundles as represented in Fig. 224; as

Fig. 233.—Very old cherry tree in Maruyaami park, Kyoto, with its limbs supported to guard against injury from winds.

bundles of limbs twenty-four to thirty inches, and sometimes four to six feet, long; and in the form of charcoal made from trunks and stems one and a half inches to six inches long, and baled in straw matting. Most of the draft animals used in Japan are either cows, bulls or stallions; at least we saw very few oxen and few geldings.

As early as 1895 the Government began definite steps looking to the improvement of horse breeding, appointing at that time a commission to devise comprehensive plans. This led to progressive steps finally culminating in 1906 in the Horse Administration Bureau, whose duties were to extend over a period of thirty years, divided into two intervals, the first, eighteen and the second, twelve years. During the first interval it is contemplated that the Government shall acquire 1,500 stallions to be distributed throughout the country for the use of private individuals, and during the second period it is the expectation that the system will have completely renovated the stock and fa-

Fig. 234.—Admiring cherry blossoms.

miliarized the people with proper methods of management so that matters may be left in their hands.

As our main purpose and limited time required undivided attention to agricultural matters, and of these to the long established practices of the people, we could give but little time to sight-seeing or even to a study of the efforts being made for the introduction of improved agricultural methods and practices. But in the very old city of Kyoto, which was the seat of the Mikado's court from be-

Fig. 235.—Entrance way to Kiyomizu temple, Kyoto.

fore 800 A. D. until 1868, we did pay a short visit to the
Kiyomizu temple, situated some three hundred yards south
from the Yaami hotel, which faces the Maruyaami park
with its centuries-old giant cherry tree, having a trunk of
more than four feet through and wide spreading branches,
now much propped up to guard against accident, as seen
in Fig. 233. These cherry trees are very extensively used
for ornamental purposes in Japan with striking effect. The
tree does not produce an edible fruit, but is very beautiful
when in full bloom, as may be seen from Fig. 234. It was
these trees that were sent by the Japanese government to
this country for use at Washington but the first lot were
destroyed because they were found to be infested and
threatened danger to native trees.

Kyoto stands amid surroundings of wonderful beauty,
the site apparently having been selected with rare acumen
for its possibilities in large landscape effects, and these
have been developed with that fullness and richness which
the greatest artists might be content to approach. We are
thinking particularly of the Kiyomizu-dera, or rather of
the marvelous beauty of tree and foliage which has overgrown it and swept far up and over the mountain summit,
leaving the temple half hidden at the base. No words,
no brush, no photographic art can transfer the effect. One
must see to feel the influence for which it was created, and
scores of people, very old and very young, nearly all Japanese, and more of them on that day from the poorer
rather than from the well-to-do class, were there, all with-
drawing reluctantly, like ourselves, looking backward, under the spell. So potent and impressive was that something from the great overshadowing beauty of the mountain, that all along up the narrow, shop-lined street leading to the gateway of the temple, seen in Fig. 235, the tini-
est bits of park effect were flourishing in the most impossible situations; and as Professor Tokito and myself
were coming away we chanced upon six little roughly
dressed lads laying out in the sand an elaborate little park,

Fig. 236.—View of Kiyomizu temple and the wooded mountain slope rising beyond, showing how dense the forest growth may become when long protected.

quite nine by twelve feet. They must have been at it
hours, for there were ponds, bridges, tiny hills and ravines
and much planting in moss and other little greens. So
intent on their task were they that we stood watching full
two minutes before our presence attracted their attention,
and yet the oldest of the group must have been under ten
years of age.

Fig. 237.—Japanese park seats at Kiyomizu-dera, Kyoto.

One partly hidden view of the temple is seen in Fig. 236,
the dense mountain verdure rising above and beyond it.
And then too, within the temple, as the peasant men and
women came before the shrine and grasped the long depending rope knocker, with the heavy knot in front of the
great gong, swinging it to strike three rings, announcing
their presence before their God, then kneeling to offer
prayers, one could not fail to realize the deep sincerity and
faith expressed in face and manner, while they were obliv-

Fig. 288.—Iris garden, Japan.

ious to all else. No Christian was ever more devout and one may well doubt if any ever arose from prayer more uplifted than these. Who need believe they did not look beyond the imagery and commune with the Eternal Spirit?

A third view of the same temple, showing resting places beneath the shade, which serve the purpose of lawn seats in our parks, is seen in Fig. 237.

Fig 239.—Street flower-vender, Japan.

That a high order of the esthetic sense is born to the Japanese people; that they are masters of the science of the beautiful; and that there are artists among them capable of effective and impressive results, is revealed in a hundred ways, and one of these is the iris garden of Fig. 238. One sees it here in the bulrushes which make the iris feel at home; in the unobtrusive semblance of a log that seems to have fallen across the run; in the hard beaten

narrow path and the sore toes of the old pine tree, telling
of the hundreds that come and go; it is seen in the dress
and pose of the ladies, and one may be sure the photog-
rapher felt all that he saw and fixed so well.

The vender of Oumi's lily that Margaret Johnson saw,
is in Fig. 239. There another is bartering for a spray of
flowers, and thus one sold the branch of red maple leaves
in our room at the Nara inn. His floral stands are borne
along the streets pendant from the usual carrying pole.

When returning to the city from the Kyoto Experiment
Station several fields of Japanese indigo were passed,
growing in water under the conditions of ordinary rice
culture, Fig. 240 being a view of one of these. The plant
is *Poligonum tinctoria,* a close relative of the smartweed.
Before the importation of aniline and alizarin dyes, which
amounted in 1907 to 160,558 pounds and 7,170,320 pounds
respectively, the cultivation of indigo was much more extensive than at present, amounting in 1897 to 160,460,000
pounds of the dried leaves; but in 1906 the production had
fallen to 58,696,000 pounds, forty-five per cent of which
was grown in the prefecture of Tokushima in the eastern
part of the island of Shikoku. The population of this prefecture is 707,565, or 4.4 people to each of the 159,450
acres of cultivated field, and yet 19,969 of these acres bore
the indigo crop, leaving more than five people to each food-
producing acre.

The plants for this crop are started in nursery beds in
February and transplanted in May, the first crop being
cut the last of June or first of July, when the fields are
again fertilized, the stubble throwing out new shoots and
yielding a second cutting the last of August or early September. A crop of barley may have preceded one of
indigo, or the indigo may be set following a crop of rice.
Such practice, with the high fertilization for every crop,
goes a long way toward supplying the necessary food. The
dense population, too, has permitted the manufacture of
the indigo as a home industry among the farmers, enabling them to exchange the spare labor of the family for

cash. The manufactured product from the reduced planting in 1907 was worth $1,304,610, forty-five per cent of
which was the output of the rural population of the prefecture of Tokushima, which they could exchange for rice
and other necessaries. The land in rice in this prefecture
in 1907 was 73,816 acres, yielding 114,380,000 pounds, or
more than 161 pounds to each man, woman and child, and
there were 65,665 acres bearing other crops. Besides this
there are 874,208 acres of mountain and hill land in the

Fig. 240.—Field of Japanese indigo, just outside the city
of Kyoto.

prefecture which supply fuel, fuel ashes and green manure
for fertilizer; run-off water for irrigation; lumber and remunerative employment for service not needed in the fields.

The journey was continued from Kyoto July 7th, taking
the route leading northeastward, skirting lake Biwa
which we came upon suddenly on emerging from a tunnel
as the train left Otani. At many places we passed water-
wheels such as that seen in Fig. 241, all similarly set, busily
turning, and usually twelve to sixteen feet in diameter
but oftenest only as many inches thick. Until we had

reached Lake Biwa the valleys were narrow with only small areas in rice. Tea plantations were common on the higher cultivated slopes, and gardens on the terraced hillsides growing vegetables of many kinds were common, often with the ground heavily mulched with straw, while the wooded or grass-covered slopes still further up showed the usual systematic periodic cutting. After passing the west end of the lake, rice fields were nearly continuous and ex-

Fig. 241.—Type of water-wheel seen very commonly on the mountain streams in Japan.

tensive. Before reaching Hachiman we crossed a stream leading into the lake but confined between levees more than twelve feet high, and we had already passed beneath two raised viaducts after leaving Kusatsu. Other crops were being grown side by side with the rice on similar lands and apparently in rotation with it, but on sharp, narrow, close ridges twelve to fourteen inches high. As we passed eastward we entered one of the important mulberry districts where the fields are graded to two levels, the higher occupied with mulberry or other crops not requiring irrigation, while the lower was devoted to rice or crops grown in rotation with it.

On the Kisogawa, at the station of the same name, there were four anchored floating water-power mills propelled by two pair of large current wheels stationed fore and aft, each pair working on a common axle from opposite sides of the mill, driven by the force of the current flowing by.

At Kisogawa we had entered the northern end of one of the largest plains of Japan, some thirty miles wide and extending forty miles southward to Owari bay. The plain has been extensively graded to two levels, the benches being usually not more than two feet above the rice paddies, and devoted to various dry land crops, including the mulberry. The soil is decidedly sandy in character but the mean yield of rice for the prefecture is 37 bushels per acre and above the average for the country at large. An analysis of the soils at the sub-experiment station north of Nagoya shows the following content of the three main plant food elements.

	Nitrogen	Phosphorus	Potassium
		Pounds per million	
		In paddy field	
Soil	1520	769	805
Subsoil	810	756	888
		In upland field	
Soil	1060	686	1162
Subsoil	510	673	1204

The green manure crops on this plain are chiefly two varieties of the "pink clover," one sowed in the fall and one about May 15th, the first yielding as high as sixteen tons green weight per acre and the other from five to eight tons.

On the plain distant from the mountain and hill land the stems of agricultural crops are largely used as fuel and the fuel ashes are applied to the fields at the rate of 10 kan per tan, or 330 pounds per acre, worth $1.20, little lime, as such, being used.

In the prefecture of Aichi, largely in this plain, with an area of cultivated land equal to about sixteen of our government townships, there is a population of 1,752,042, or a density of 4.7 per acre, and the number of households of farmers was placed at 211,033, thus giving to each farmer's family an average of 1.75 acres, their chief industries being rice and silk culture.

Soon after leaving the Agricultural Experiment Station of Aichi prefecture at An Jo we crossed the large Yahagigawa, flowing between strong levees above the level of the rice fields. Mulberries, with burdock and other vegetables were growing upon all of the tables raised one to two feet above the rice paddies, and these features continued past Okasaki, Koda and Kamagori, where the hills in many places had been recently cut clean of the low forest growth and where we passed many large stacks of pine boughs tied in bundles for fuel. After passing Goyu sixty-five miles east from Nagoya, mulberry was the chief crop. Then came a plain country which had been graded and leveled at great cost of labor, the benches with their square shoulders standing three to four feet above the paddy fields; and after passing Toyohashi some distance we were surprised to cross a rather wide section of comparatively level land overgrown with pine and herbaceous plants which had evidently been cut and recut many times. Beyond Futagawa rice fields were laid out on what appeared to be similar land but with soil a little finer in texture, and still further along were other flat areas not cultivated.

At Maisaka quite half the cultivated fields appear to be in mulberry with ponds of lotus plants in low places, while at Hamamatsu the rice fields are interspersed with many square-shouldered tables raised three to four feet and occupied with mulberry or vegetables. As we passed upon the flood plain of the Tenryugawa, with its nearly dry bed of coarse gravel half a mile wide, the dwellings of farm villages were many of them surrounded with nearly solid, flat-topped, trimmed evergreen hedges nine to twelve feet high, of the umbrella pine, forming beautiful and effective screens.

At Nakaidzumi we had left the mulberry orchards for those of tea, rice still holding wherever paddies could be formed. Here, too, we met the first fields of tobacco, and at Fukuroi and Homouchi large quantities of imported Manchurian bean cake were stacked about the station, having evidently been brought by rail. At Kanaya we passed through a long tunnel and were in the valley of the Oigawa, crossing the broad, nearly dry stream over a bridge of nineteen long spans and were then in the prefecture of Shizuoka where large fields of tea spread far up the hillsides, covering extensive areas, but after passing the next station, and for seventeen miles before reaching Shizuoka we traversed a level stretch of nearly continuous rice fields.

The Shizuoka Experiment Station is devoting special attention to the interests of horticulture, and progress has already been made in introducing new fruits of better quality and in improving the native varieties. The native pears and peaches, as we found them served on the hotel tables in either China or Japan, were not particularly attractive in either texture or flavor, but we were here permitted to test samples of three varieties of ripe figs of fine flavor and texture, one of them as large as a good sized pear. Three varieties of fine peaches were also shown, one unusually large and with delicate deep rose tint, including the flesh. If such peaches could be canned so as to retain their delicate color they would prove very attractive for the table. The flavor and texture of this peach were also excellent, as was the case with two varieties of pears.

The station was also experimenting with the production of marmalades and we tasted three very excellent brands, two of them lacking the bitter flavor. It would appear that in Japan, Korea and China there should be a very bright future along the lines of horticultural development, leading to the utilization of the extensive hill lands of these countries and the development of a very extensive export trade, both in fresh fruits and marmalades, preserves and the canned forms. They have favorable clima-

Fig. 242.—Views of buildings and grounds at the Shizuoka Experiment Station.

tic and soil conditions and great numbers of people with temperament and habits well suited to the industries, as well as an enormous home need which should be met, in addition to the large possibilities in the direction of a most profitable export trade which would increase opportunities for labor and bring needed revenue to the people.   In Fig. 242 are three views at this station, the lower showing a steep terraced hillside set with oranges and other fruits, holding out a bright promise for the future.

Peach orchards were here set on the hill lands, the trees six feet apart each way.   They come into bearing in three years, remain productive ten to fifteen years, and the returns are 50 to 60 yen per tan, or at the rate of $100 to $120 per acre.   The usual fertilizers for a peach orchard are the manure-earth-compost, applied at the rate of 3300 pounds per acre, and fish guano applied in rotation and at the same rate.

Shizuoka is one of the large prefectures, having a total area of 3029 square miles; 2090 of which are in forest; 438 in pasture and *genya* land, and 501 square miles cultivated, not quite one-half of which is in paddy fields. The mean yield of paddy rice is nearly 33 bushels per acre. The prefecture has a population of 1,293,470, or about four to the acre of cultivated field, and the total crop of rice is such as to provide 236 pounds to each person.

At many places along the way as we left Shizuoka July 10th for Tokyo, farmers were sowing broadcast, on the water, over their rice fields, some pulverized fertilizer, possibly bean cake.   Near the railway station of Fuji, and after crossing the boulder gravel bed of the Fujikawa which was a full quarter of a mile wide, we were traversing a broad plain of rice paddies with their raised tables, but on them pear orchards were growing, trained to their overhead trellises.   About Suduzuka grass was being cut with sickles along the canal dikes for use as green manure in the rice fields, which on the left of the railway, stretched eastward more than six miles to beyond Hara where we passed

27

into a tract of dry land crops consisting of mulberry, tea
and various vegetables, with more or less of dry land rice,
but we returned to the paddy land again at Numazu, in
another four miles. Here there were four carloads of beef
cattle destined for Tokyo or Yokohama, the first we had
seen.

It was at this station that the railway turns northward
to skirt the eastern flank of the beautiful Fuji-yama, rising

Fig. 243.—Japanese ladies eating buckwheat macaroni with chopsticks.

to higher lands of a brown loamy character, showing many
large boulders two feet in diameter. Horses were here
moving along the roadways under large saddle loads of
green grass, going to the paddy fields from the hills, which
in this section are quite free from all but herbaceous
growth, well covered and green. Considerable areas were
growing maize and buckwheat, the latter being ground into flour and made into macaroni which is eaten with chopsticks, Fig. 243, and used to give variety to the diet of rice
and naked barley. At Gotenba, where tourists leave the

train to ascend Fuji-yama, the road turns eastward again
and descends rapidly through many tunnels, crossing the
wide gravelly channel of the Sakawagawa, then carrying
but little water, like all of the other main streams we had
crossed, although we were in the rainy season. This was
partly because the season was yet not far advanced; partly
because so much water was being taken upon the rice
fields, and again because the drainage is so rapid down
the steep slopes and comparatively short water courses.
Beyond Yamakita the railway again led along a broad
plain set in paddy rice and the hill slopes were terraced
and cultivated nearly to their summits.

Swinging strongly southeastward, the coast was reached
at Noduz in a hilly country producing chiefly vegetables,
mulberry and tobacco, the latter crop being extensively
grown eastward nearly to Oiso, beyond which, after a mile
of sweet potatoes, squash and cucumbers, there were paddy
fields of rice in a flat plain. Before Hiratsuka was
reached the rice paddies were left and the train was crossing a comparatively flat country with a sandy, sometimes
gravelly, soil where mulberries, peaches, eggplants, sweet
potatoes and dry land rice were interspersed with areas
still occupied with small pine and herbaceous growth or
where small pine had been recently set. Similar conditions prevailed after we had crossed the broad channel
of the Banyugawa and well toward and beyond Fujishiwa
where a leveled plain has its tables scattered among the
fields of paddy rice, this being the southwest margin of
the Tokyo plain, the largest in Japan, lying in five prefectures, whose aggregate area of 1,739,200 acres of arable
lands was worked by 657,235 families of farmers; 661,613
acres of which was in paddy rice, producing annually
some 19,198,000 bushels, or 161 pounds for each of the
7,194,045 men, women and children in the five prefectures,
1,818,655 of whom were in the capital city, Tokyo.

Three views taken in the eastern portion of this plain
in the prefecture of Chiba, July 17th, are seen in Fig.

Fig. 244.—Three landscapes in the Tokyo plain, the upper two largely in sweet potatoes, following wheat, the lower in peanuts.

Fig. 245.—Two methods of utilizing coarse straw and litter for mulching and fertilizing at the same time.

244, in two of which shocks of wheat were still standing in the fields among the growing crops, badly weathered and the grain sprouting as the result of the rainy season. Peanuts, sweet potatoes and millet were the main dry land crops then on the ground, with paddy rice in the flooded basins. Windsor beans, rape, wheat and barley had been harvested. One family with whom we talked were threshing their wheat. The crop had been a good one and was yielding between 38.5 and 41.3 bushels per acre, worth at the time $35 to $40. On the same land this farmer secures a yield of 352 to 361 bushels of potatoes, which at the market price at that time would give a gross earning of $64 to $66 per acre.

Reference has been made to the extensive use of straw in the cultural methods of the Japanese. This is notably the case in their truck garden work, and two phases of this are shown in Fig. 245. In the lower section of the illustration the garden has been ridged and furrowed for transplanting, the sets have been laid and the roots covered with a little soil; then, in the middle section, showing the next step in the method, a layer of straw has been pressed firmly above the roots, and in the final step this would be covered with earth. Adopting this method the straw is so placed that (1) it acts as an effective mulch without in any way interfering with the capillary rise of water to the roots of the sets; (2) it gives deep, thorough aeration of the soil, at the same time allowing rains to penetrate quickly, drawing the air after it; (3) the ash ingredients carried in the straw are leached directly to the roots where they are needed; (4) and finally the straw and soil constitute a compost where the rapid decay liberates plant food gradually and in the place where it will be most readily available. The upper section of the illustration shows rows of eggplants very heavily mulched with coarse straw, the quantity being sufficient to act as a most effective mulch, to largely prevent the development of weeds and to serve during the rainy season as a very material fertilizer.

In growing such dry land crops as barley, beans, buckwheat or dry land rice the soil of the field is at first fitted
by plowing or spading, then furrowed deeply where the
rows are to be planted. Into these furrows fertilizer is
placed and covered with a layer of earth upon which the
seed is planted. When the crop is up, if a second fertilization is desired, a furrow may be made alongside each

Fig. 246.—Section of soil study field, Imperial Agricultural Experiment Station,
Tokyo, Japan.

row, into which the fertilizer is sowed and then covered.
When the crop is so far matured that a second may be
planted, a new furrow is made, either midway between two
others or adjacent to one of them, fertilizer applied and
covered with a layer of soil and the seed planted. In
this way the least time possible is lost during the growing
season, all of the soil of the field doing duty in crop production.

It was our privilege to visit the Imperial Agricultural
Experiment Station at Nishigahara, near Tokyo, which

is charged with the leadership of the general and technical agricultural research work for the Empire. The work is divided into the sections of agriculture, agricultural chemistry, entomology, vegetable pathology, tobacco, horticulture, stock breeding, soils, and tea manufacture, each with their laboratory equipment and research staff, while the forty-one prefectural stations and fourteen sub-stations are charged with the duty of handling all specific local, practical problems and with testing out and applying conclusions and methods suggested by the results obtained at the central station, together with the local dissemination of knowledge among the farmers of the respective prefectures.

A comprehensive soil survey of the arable lands of the Empire has been in progress since before 1893, excellent maps being issued on a scale of 1 to 100,000, or about 1.57 inch to the mile, showing the geological formations in eight colors with subdivisions indicated by letters. Some eleven soil types are recognized, based on physical composition and the areas occupied by these are shown by means of lines and dots in black printed over the colors. Typical profiles of the soil to depths of three meters are printed as insets on each sheet and localities where these apply are indicated by corresponding numbers in red on the map.

Elaborate chemical and physical studies are also being made in the laboratories of samples of both soil and subsoil. The Imperial Agricultural Experiment Station is well equipped for investigation work along many lines and that for soils is notably strong. In Fig. 246 may be seen a portion of the large immersed cylinders which are filled with typical soils from different parts of the Empire, and Fig. 247 shows a portion of another part of their elaborate outfit for soil studies which are in progress.

It is found that nearly all cultivated soils of Japan are acid to litmus, and this they are inclined to attribute to the presence of acid hydro-aluminum silicates.

The Island Empire of Japan stretches along the Asiatic coast through more than twenty-nine degrees of latitude from the southern extremity of Formosa northward to the middle of Saghalin, some 2300 statute miles; or from the latitude of middle Cuba to that of north Newfoundland and Winnipeg; but the total land area is only 175,428 square miles, and less than that of the three states of Wisconsin, Iowa and Minnesota. Of this total land area only 23,698 square miles are at present cultivated; 7151 square miles in the three main islands are weed and pasture land. Less than fourteen per cent of the entire land area is at present under cultivation.

If all lands having a slope of less than fifteen degrees may be tilled, there yet remain in the four main islands, 15,400 square miles to bring under cultivation, which is an addition of 65.4 per cent to the land already cultivated.

In 1907 there were in the Empire some 5,814,362 households of farmers tilling 15,201,969 acres and feeding 3,522,877 additional households, or 51,742,398 people. This is an average of 3.4 people to the acre of cultivated land, each farmer's household tilling an average of 2.6 acres.

The lands yet to be reclaimed are being put under cultivation rapidly, the amount improved in 1907 being 64,448 acres. If the new lands to be reclaimed can be made as productive as those now in use there should be opportunity for an increase in population to the extent of about 35,000,000 without changing the present ratio of 3.4 people to the acre of cultivated land.

While the remaining lands to be reclaimed are not as inherently productive as those now in use, improvements in management will more than compensate for this, and the Empire is certain to quite double its present maintenance capacity and provide for at least a hundred million people with many more comforts of home and more satisfaction for the common people than they now enjoy.

Since 1872 there has been an increase in the population of Japan amounting to an annual average of about 1.1 per cent, and if this rate is maintained the one hundred million mark would be passed in less than sixty years. It

Fig. 247.—Part of equipment for chemical soil studies, Imperial Agricultural Experiment Station, Tokyo, Japan.

appears probable however that the increased acreage put under cultivation and pasturage combined, will more than keep pace with the population up to this limit, while the improvement in methods and crops will readily permit a second like increment to her population, bringing that for the present Empire up to 150 millions. Against this view, perhaps, is the fact that the rice crop of the twenty years ending in 1906 is only thirty-three per cent greater than the crop of 1838.

In Japan, as in the United States, there has been a strong movement from the country to the city as a natural result of the large increase in manufactures and commerce, and the small amount of land per each farmer's household. In 1903 only .23 per cent of the population of Japan were living in villages of less than 500, while 79.06 per cent were in towns and villages of less than 10,000 people, 20.7 per cent living in those larger. But in 1894 84.36 per cent of the population were living in towns and villages of less than 10,000, and only 15.64 per cent were in cities, towns and villages of over 10,000 people; and while during these ten years the rural population had increased at the rate of 640 per 10,000, in cities the increase had been 6,174 per 10,000.

Japan has been and still is essentially an agricultural nation and in 1906 there were 3,872,105 farmers' households, whose chief work was farming, and 1,581,204 others whose subsidiary work was farming, or 60.2 per cent of the entire number of households. A like ratio holds in Formosa. Wealthy land owners who do not till their own fields are not included.

Of the farmers in Japan some 33.34 per cent own and work their land. Those having smaller holdings, who rent additional land, make up 46.03 per cent of the total farmers; while 20.63 per cent are tenants who work 44.1 per cent of the land. In 1892 only one per cent of the land holders owned more than twenty-five acres each; those holding between twenty-five acres and five acres made up 11.7 per cent; while 87.3 per cent held less than five acres each. A man owning seventy-five acres of land in Japan is counted among the "great land-holders". It is never true, however, except in the Hokkaido, which is a new country agriculturally, that such holdings lie in one body.

Statistics published in "Agriculture in Japan", by the Agricultural Bureau, Department of Agriculture and Commerce, permit the following statements of rent, crop returns, taxes and expenses, to be made. The wealthy land owners who rent their lands receive returns like these:

	For paddy field, per acre.	For upland field, per acre.
Rent	$27.98	$13.53
Taxes	7.34	1.98
Expenses	1.72	2.48
Total expenses	$9.06	$4.46
Net profit	18.92	9.07

It is stated, in connection with these statistics, that the rate of profit for land capital is 5.6 per cent for the paddy field, and 5.7 per cent for the upland field. This makes the valuation of the land about $338 and $159 per acre, respectively. A land holder who owns and rents ten acres of paddy field and ten acres of upland field would, at these rates, realize a net annual income of $279.90.

Peasant farmers who own and work their lands receive per acre an income as follows:

	For paddy field, per acre.	For upland field, per acre.
Crop returns	$55.00	$30.72
Taxes	7.34	1.98
Labor and expenses	36.20	24.00
Total expense	$43.54	$25.98
Net profit	11.46	4.74

The peasant farmer who owns and works five acres, 2.5 of paddy and 2.5 of upland field, would realize a total net income of $40.50. This is after deducting the price of his labor. With that included, his income would be something like $91.

Tenant farmers who work some 41 per cent of the farm lands of Japan, would have accounts something as follows:

	For paddy field, 1 crop. per acre.	2 crops.	For upland field, per acre.
Crop returns	$49.03	$78.62	$41.36
Tenant fee	23.89	31.58	13.52
Labor	15.78	25.79	14.69
Fertilization	7.82	17.30	10.22
Seed	.82	1.40	1.57
Other expenses	1.69	2.82	1.66
Total expenses	$50.00	$78.89	$41.66
Net profit	—.97	—.27	—.30

This statement indicates that tenant farmers do not realize enough from the crops to quite cover expenses and the price named for their labor. If the tenant were renting five acres, equally divided between paddy and upland field, the earning would be $73.00 or $99.73 according as one or two crops are taken from the paddy field, this representing what he realizes on his labor, his other expenses absorbing the balance of the crop value.

But the average area tilled by each Japanese farmer's household is only 2.6 acres, hence the average earning of the tenant household would be $37.95 or $51.86. A clearer view of the difference in the present condition of farmers in Japan and of those in the United States may be gained by making the Japanese statement on the basis of our 160-acre farm, as expressed in the table below:

	For paddy field. For 80 acres.	For upland field. For 80 acres.	Total. 160 acres.
Crop returns	$4,400.00	$2,457.60	$6,857.60
Taxes	$587.20	$158.40	$745.60
Expenses	1,633.60	744.80	2,378.40
Labor	1,262.40	1,175.20	2,437.60
Total cost	$3,488.20	$2,078 40	$5,561 60
Net return	916.80	379.20	1,296.00
Return including labor	2,179.20	1,554.40	3,733.60

In the United States the 160-acre farm is managed by and supports a single family, but in Japan, as the average household works but 2.6 acres, the earnings of the 160 acres are distributed among some 61 household, making the net return to each but $21.25, instead of $1296, and including the labor as earning, the income would be $39.96 more, or $60.67 per household instead of $3733.60, the total for a 160-acre farm worked under Japanese conditions.

These figures reveal something of the tense strain and of the terrible burden which is being carried by these people, over and above that required for the maintenance of the household. The tenant who raises one crop of rice pays a rental of $23.89 per acre. If he raises two crops he pays $31.58; if it is upland field, he pays $13.52. To these amounts he adds $10.33, $21.52 or $13.45 respectively for fertilizer, seed and other expenses, making a total investment of $34.22, $53.10 or $26.97 per acre, which would require as many bushels of wheat sold at a dollar a bushel to cover this cost. In addition to this he assumes all the risks of loss from weather, from insects and from blight, in the hope that he may recoup his expenses and in addition have for his services $14.81, $25.52 or $14.39 for the season's work.

The burdens of society, which have been and still are so largely burdens of war and of government, with all nations, are reflected with almost blinding effect in the land taxes of Japan, which range from $1.98, on the upland, to $7.34 per acre on the paddy fields, making a quarter section, without buildings, carry a burden of $300 to $1100 annually. Japan's budget in 1907 was $134,941,113, which is at the rate of $2.60 for each man, woman and child; $8.90 for each acre of cultivated land, and $23, for each household in the Empire. When such is the case it is not strange that scenes like Fig. 248 are common in Japan today where, after seventy years, toil may not cease.

There is a bright, as well as a pathetic side to scenes like this. The two have shared for fifty years, but if the days have been full of toil, with them have come strength of body, of mind and sterling character. If the burdens have been heavy, each has made the other's lighter, the satisfaction fuller, the joys keener, the sorrows less difficult to bear; and the children who came into the home and have gone from it to perpetuate new ones, could not

Fig. 248.—After seventy years, toil may not cease.

well be other than such as to contribute to the foundations of nations of great strength and long endurance.

Reference has been made to the large amount of work carried on in the farmers' households by the women and children, and by the men when they are not otherwise employed, and the earnings of this subsidiary work have materially helped to piece out the meagre income and to meet the relatively high taxes and rent.

INDEX.

A

Acidity of soils, 424.

Acres per capita, U. S. 1; Orient, 1, 2, 193, 410, 425.

Afforestation, 151, 155, 156, 159, 398; tract, 217-220.

Agricultural college, 381.

Aichi, 413.

Akashi Experiment Station, 22, 396.

Amur river, 351.

Analysis, ashes, 207; compost, 211; excreta, 194; genya, 211; mil⸗, 149; soil, 413.

Angleworms, 205.

Animal diet, 135.

Antung, 358, 365.

Area, cultivated land, 6, 425; per family, 425, 429; of gardens, 377; of rice paddies, 277.

Area, Aichi, 414; Japan, 425; Nara, 400; Shantung, 216; Shizuoka, 417; Tokushima, 417; Tokyo plain, 419.

Area, forests, 160; genya, 209; legumes, 213; rice fields, 7, 8, 27, 271, 272; tea, 326; textiles, 164; wheat, 272; rush, 395.

Ashes as fertilizer, 9, 68, 169, 182, 207, 251, 286, 296, 380, 392, 410, 425.

Astragalus sinicus, 10, 380.

B

Bags, of matting, 159, 165, 305, 306, 308, 359; paper, 396.

Bamboo, 62, 64, 127, 130, 132, 133, 138, 147, 157, 165, 188, 290, 293, 301, 385, 387, 388; sprouts, 131.

Bananas, 82.

Barley, 6, 53, 54, 226, 227, 242, 267, 272, 306, 309, 329, 362, 379, 410, 422, 423; tying stems, 206.

Beans, 33, 122, 169, 213, 226, 255, 290, 305, 309, 319, 329, 380, 422, 423; sprouted, 134.

Bean cakes, 257, 378, 392, 417; export, 357, 358, 415.

Bean curd, 143.

Beauty of landscapes, 389-392, 400, 405, 409.

Beds, chimney, 141, 142, 248.

Beef cattle, 418.

Beggars, 121, 176.

Bellows, 143.

Bending wood, 89.

Bibo, 396.

Birds, 62.

Blumann, Dr. John, 93, 96.

Boats, 77, 78, 83, 86, 171.

Bombyx mori, 319.

Borrowing money, 152.

Bound feet, 62, 122, 230.

Bow, for whipping cotton, 127.

Bow-brace, 87.

Boxer uprising, 217.

Braid, straw, 165, 226.

Braziers, 138.

Brick, 142, 143, 162, 163.

Brick vaults, 50, 52.

Bridge building, 299.

Bucket and well sweep, 297, 299.

Buckwheat, 359, 418, 423.

Buffalo, water, 145, 149, 150, 235, 335, 336, 338.

Buffalo-horn, nut, 134.

Building materials, 160-163, 232, 285, 358, 337, 395.